MW00989107

Dynamic Systems for Everyone

Asish Ghosh

Dynamic Systems for Everyone

Understanding How Our World Works

Second Edition

Asish Ghosh
Plymouth, MA, USA

ISBN 978-3-319-43942-6 ISBN 978-3-319-43943-3 (eBook)
DOI 10.1007/978-3-319-43943-3

Library of Congress Control Number: 2016952253

Printed on acid-free paper

This Springer imprint is published by Springer Nature
The registered company is Springer International Publishing AG
The registered company address is: Gewerbestrasse 11, 6330 Cham, Switzerland

To: Ahan, Ashapurna, Annapurna,
Narasimha, and Owen

Foreword

Having spent over a half century in controlling industrial processes, as has the author, I too believe that the understanding we have gained could enrich the totality of human understanding and evolution.

The approach of a process control person is to first gain a full understanding of the nature of a process before attempting to control it. We must clearly understand if it is a batch or a continuous process; we must know its inertia, time constants, safety aspects, and the like. Once we understand its full "personality" and determine the goal we want to reach by controlling it, we can figure out what we can manipulate to "herd" that process in the right direction and what we need to measure, in order to make sure that our process will reach its target safely and on time.

For example, once we clearly understand that the goal of controlling a nuclear power plant is to generate electricity while making it impossible for it to blow up, we also know that we need a totally reliable energy source to prevent its meltdown, a source that nothing on this planet can turn off. Understanding that, it is easy to conclude that such an energy source is gravity, and therefore we should place the reactor underwater.

Similarly, life on this planet is a process fuelled by the energy of the Sun, and life is maintained so long as the activity of plant and animal life is in balance. Plants consume carbon dioxide while generating oxygen, which is then used by the animals or man who exhale carbon dioxide. We know that the balance is upset when the carbon dioxide concentration in the air rises. During my lifetime, the human population on the planet has increased from 2.1 to over 7 billion, and the carbon dioxide concentration of the air has also tripled. Now, if the goal of this process control system is to maintain human life on Earth, it is pretty obvious that we must satisfy our increased energy needs from a carbon-free and inexhaustible source, namely, the Sun.

When I preach about technical stuff, such as the potentials of process control in such areas to my wife or kids, their eyes get blurred and these windows of their brains close. It is for this reason that books are needed which convince the nontechnical reader that the scientific principles of process control are applicable to all systems, including biological and social ones, and do it in such a way that their eyes do not get blurred.

Asish Ghosh has written such a book. Read it!

Stamford, CT, USA Béla Lipták

Preface

After spending many years as a systems engineer for controlling various production processes, I began to view the world more and more as a collection of systems. In my engineering work, systems existed as manufacturing entities, such as chemical plants, oil refineries, food and drug manufacturing, and materials handling systems. As my systems view developed beyond my work, I came to recognize other types of systems, such as living beings, like humans, dogs, and elephants, and mechanical entities like motorcars, bicycles, and robots.

A system may be variously defined as "a complex whole, a set of connected things or parts, and an organized body of material or immaterial things." Various organizations and institutions created by human beings, such as banks, schools, and businesses, are also systems or may be considered as subsystems of larger systems. Thus, the definition of a system is not confined to any particular discipline but is widely applicable to many different fields.

This book is a study of the behavior of these systems while they interact with other systems and with their environment. It explains the basic principles of various mechanical and electromechanical systems and then guides the reader to understand how the same rules may apply to social, political, and economic systems, as well as to everyday life. It also explains how by using systems principles we can reduce or eliminate many unintended consequences of our actions.

Today, we are facing major technological and social challenges, such as global warming, nuclear catastrophe, cyber security, worldwide shortage of potable water, and violent religious extremisms. Addressing these problems will not only require knowledge in technologies and social sciences but also a deep understanding of the systems and their behaviors.

The book is meant for general readers who can apply systems principles in their work and in everyday life. Social scientists, healthcare providers, economists, and specialists in many other disciplines will find that increased system knowledge can be of great help in their professional work. Control system engineers will also benefit from reading this book by getting a better understanding of how feedback control, with which they are so familiar, may also be applied in natural and social contexts. Given that this book is intended for a broad audience, including nontechnical readers, I have tried to avoid mathematical formulas and technical jargons as far as possible.

Chapter 1 provides an overview of system thinking with simple examples of diverse types of systems and their underlying similarities. It discusses why feedback is fundamental to the understanding of the behavior of a system. It classifies systems into various types, such as natural, social, political, economic, and engineered, and discusses why information exchange is fundamental to a system's behavior.

Chapter 2 gives the early history of engineered systems including descriptions of the working of the Watt governor, which was developed to control steam engines. It describes control loops and their various parts, such as sensors, actuators, and controllers. It discusses the importance of getting feedback at the right time and how delays can adversely affect the behavior of a system.

Chapter 3 covers non-engineered systems, such as political, social, and biological entities. It shows how causal loop diagrams can be used to depict such systems, and discusses feedback and biofeedback, along with dead time and lag, as they apply in these systems. It documents a case study on the dynamics of youth violence.

Chapter 4 describes the various behavior patterns of systems, such as growth, decay, oscillation, and goal seeking. It discusses linear and nonlinear behaviors and compares detail complexity with dynamic complexity. It explains why system behaviors are sometimes so unpredictable.

Chapter 5 discusses the importance of modeling and simulation for the understanding of a system's behavior. It describes software packages that are available for modeling and simulation. It builds a model of the youth violence mentioned earlier and discusses the many different uses of models.

Chapter 6 outlines the problems one faces in optimizing a system, which has multiple goals and constraints. It discusses methods of optimization, such as using models, hill climbing techniques, and kaizen (optimization in small steps). It details a procedure for optimizing a decision-making process.

Chapter 7 discusses why a distributed structure leads to a more robust system. It describes the need for a decentralized structure, as systems get increasingly complex. Finally, it discusses agent-based modeling and simulation techniques.

Discrete events and procedures are part of any system, but they have not been discussed in any length in many system studies. Chapter 8 makes readers aware of their effects on system behavior. It depicts loop diagrams involving discrete events and procedures and show ways to model them.

Any action to manipulate a system may produce desired results but also may lead to unintended consequences. Chapter 9 illustrates a number of unintended consequences and shows how they may be reduced by increasing system awareness.

The final chapter (Chap. 10) highlights the seven main traits of a person that takes the system approach. It starts with a discussion on the creation of the right mental and conceptual frameworks and then on building realistic and interactive models. That is followed by discussions on optimization, improving efficiency and robustness, and making improvements in small steps. It ends with a discussion on a holistic worldview.

The world around us is getting more complex, and these changes are affecting all human societies and are creating new technological and social problems. The Epilogue includes a brief discussion on the role of system science in meeting these challenges.

As a reader, you may follow these chapters sequentially, or for a quick overview, read Chaps. 1 and 10 first and then delve into the others. If you are mainly interested in biological and social sciences, you may read Chaps. 1, 3, 5, 6, 9, and 10. The appendices elaborate some of the points made in these chapters.

The book includes many examples, some of them are rather simplistic and obvious; however, they are useful because they illustrate many of the basic systems principles. Then there are those that seem to be too complicated; they may be skipped during the first reading, keeping them for later consideration.

Readers are strongly urged to "get their hands dirty" by modeling the simple systems outlined in this book or by modifying one of those available at the websites of modeling software suppliers. That will allow observing the behavior of those systems under different operating conditions and thus gain better appreciation of the concepts outlined before embarking on projects that are more ambitious.

Plymouth, MA, USA Asish Ghosh

Acknowledgments

I am very grateful to a number of my former colleagues, friends, and relations who helped me in writing this book.

Lynn Craig, Howard Rosenof, and Ray Schlunk helped me by painstakingly reviewing and editing most of the chapters. Gene Bellinger's words of wisdom and reviews helped me shape the narratives. At the very early stage of my writing, Alex Pirie gave me a set of valuable feedbacks and suggestions. Sharon Lim-Hing offered feedback for the first few chapters.

Peter Martin, a former colleague and friend, provided information on dynamic performance measures. Sam Kim, Jin Min Lee, and Erik Nordbye provided a valuable set of information on the youth violence project conducted by Emmanuel Gospel Center in Boston. George Karam with his philosophical insights provided me with a valuable set of suggestions.

My cousin Ashit Sarkar and my two daughters Anna and Asha helped me with the structure and the contents, while Narasimha and Owen gave me valuable sets of suggestions. Last but not least, my dear friend Margaret painstakingly went through the first and the final chapters to make them more readable.

I am thankful to Betsy Hallett, Gabriel Dannunzio, Ashok Vichare, and Joanne Jordan for their help in updating the book for the second edition.

About the Author

Asish Ghosh is a Control Systems Engineer with over 40 years of professional experience. He has held various research, engineering, and consulting positions while working for ICI in England, The Foxboro Company, and ARC Advisory Group in Massachusetts, USA. He has presented and published numerous papers, and is the joint author of the first book on batch process control. He has lectured extensively and has taught courses in the USA and in Europe. He was a member of the ISA/SP88 and IEC (International Electrotechnical Commission) where he was actively involved in the generation of control system standards. He studied Physics in Delhi University, India and Control Engineering in Cambridge University, England. He is a Chartered Engineer registered in UK. He now resides in the historic town of Plymouth in Massachusetts with his dog Oskar.

Contents

1

Thinking in Systems

Abstract Systems are all around us, they are sets of interrelated objects or entities that interact with each other. They can be living beings, such as humans, animals, and plants. They can be mechanical entities, such as automobiles, ships, and airplanes; or industrial plants, such as oil refineries, chemical plants, and electric power generators. There are also other entities, such as social, political, and business systems. The behaviors of these systems are shaped by their environments, by the actions and interactions of their own subentities, and by human beings. However, it is interesting to note that all these disparate systems exhibit some common behavior patterns.

This chapter introduces basic system concept by describing a number of different systems in various disciplines. It discusses the importance of information exchange and feedback in shaping system behavior. How system knowledge can help us understand and cope with them and enable us to build or modify them to address our needs and desires. Finally, the chapter introduces the concept of system view.

1.1 Global Warming

We are now well aware of global warming and its negative consequences. Most scientists seem to agree that an increase in carbon dioxide (CO_2) gas in earth's atmosphere is the primary reason for this climate change. CO_2 comes from carbon, which is abundant in nature. Coal, graphite, and diamond are solid carbons, while petroleum is a liquid or viscous compound of carbon. A very large part of all living beings is made of carbon compounds. Carbon

A. Ghosh, *Dynamic Systems for Everyone*,
DOI 10.1007/978-3-319-43943-3_1

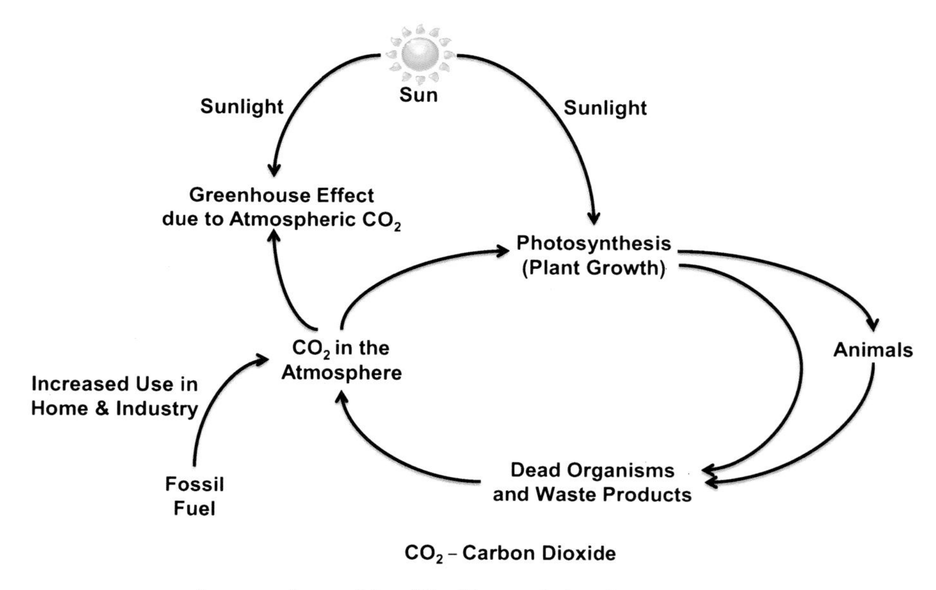

Fig. 1.1 The carbon cycle

combines with oxygen, which is abundant in earth's atmosphere, by various natural and human activated processes to form CO_2, which goes back to the atmosphere.

Plants use CO_2 and sunlight to make their own food in a process known as photosynthesis (Fig. 1.1). Herbivores get carbon by eating vegetable matters, while omnivores and carnivores get it from plants and other animals. When plants and animals die and decay the carbon generally goes back to the atmosphere as CO_2. However, they may be converted into fossil fuel, like coal and oil, if buried in earth under the right circumstances, but that can take millions of years. These interactions are collectively termed as the carbon cycle, which has been occurring in nature for a very long time.

Atmospheric CO_2 level remained steady when the world population was small and human beings used renewable resources such as wood and cow dung for cooking and heating. The balance was altered when we began using fossil fuels, such as coal or gasoline in large quantities for heating and as a source of energy for automobiles and electric power plants. Most of the burnt carbon now enters the atmosphere as CO_2.

Since the Industrial Revolution in 1700s, burning of oil, coal, and gas, combined with deforestation, has increased CO_2 concentrations in the atmosphere. In 2005, the global atmospheric concentration of CO_2 was found to be 35 % higher than it was before the Industrial Revolution. CO_2 is called a greenhouse gas because it traps heat from solar radiation in the atmosphere. Excess of greenhouse gas is the leading cause of global warming and the melting of ice caps in the Polar Regions.

Various ideas have been put forward to counter the increase of CO_2 in the atmosphere, for example, carbon sequestering where CO_2 gas is injected to abandoned mines or deep in the ocean, but they have yet to be tried in any large scale. The immediate and effective solution would be to reduce the use of fossil fuel and switch to green or renewable energy, such as solar, wind, and tidal powers. However, that involves economic and political systems, both national and international, which are quite complex.

Carbon cycle is an example of an interconnected system, where various natural, biological, and social factors interact to produce results that are often detrimental to human society. By looking at a problem like this from a system perspective, we are able to get a wider and more complete grasp of the various factors that are causing it and can formulate better and more long-term solutions.

1.2 What Is a System?

We all have some notion of what a system is but often find it difficult to define it precisely. The Oxford English Dictionary specifies a system in three ways—"A complex whole; a set of connected things or parts; an organized body of material or immaterial things." The three words: complex, connected, and organized, aptly describe the essential characteristics of a system.

Systems are all around us. They may be large or small and are composed of subsystems or components, which dynamically connect with each other to exhibit complex behavior. These systems and their components may be living beings such as humans, dogs, or elephants, or mechanical entities, such as motorcars, bicycles, or airplanes. They may also be manufacturing entities, such as chemical plants, oil refineries, electric power plants, or component manufacturing plants. Various organizations and institutions created by human beings, such as banks, schools, and business units are systems themselves, and may also be considered as components or subsystems of larger systems. For example, a school system in a town is a part of a nationwide school system whereas a bank is a part of an economic system or the international banking system. A component in a system can be quite complex and may itself be considered as a system. Thus, an automobile may be considered as a component of a system, where the driver and the car together constitute a system. However, an automobile by itself may be considered a system where its engine, brakes, and wheels are components or subsystems.

System science is the study of the dynamic behavior of systems, based on the interactions of its components and their interactions with other such systems. These interactions could be between humans, between animals, between machines, or between organizations. There could also be interactions between components of different types, such as between human beings and animals (dogs and cats), between human beings and machines (automobiles and motorbikes), and between human beings and institutions (banks, schools, and places of employment).

Each discipline has its own way of interacting and uses its own jargon (terminology) but there are also commonalities of behavior between these systems. There are interesting parallels between a person trying to drive an automobile, a teacher trying to improve the grade of a student, or a plant manager trying to optimize a production process. System theories were developed for better control and optimization of chemical and mechanical production processes. These same theories if applied judiciously may produce benefits in other fields, such as politics, law, sociology, psychology, medicine, and marketing.

1.3 Examples of Systems in Real Life

1.3.1 A Home Heating System

Central heating in your home may be considered a simple system, where you may set the thermostat to a certain value and the system is supposed to maintain that temperature (Fig. 1.2). Let us assume that your heating system is fired by oil or gas, which heats a room by circulating hot water or blowing hot air. The thermostat allows you to set the desired temperature and it has a built-in sensor that reads the room temperature. The thermostat continuously compares the set temperature (set point) with the room temperature and does nothing when they are nearly equal. If the temperature in the room falls below the set temperature by a preset amount, usually by a degree, it sends a signal to the furnace to switch on.

As the furnace switches on, it heats the circulating water or air and thus heats up the room. When the temperature gets hotter than the set point, the thermostat sends a signal to the furnace to shut it off to stop any further rise in temperature. This is an example of a simple on/off control, which maintains a reasonably constant temperature in the room.

It is important to observe here that the thermostat cannot do its job without continuously monitoring the variable that it wants to control. The thermostat senses the temperature in the room and the mechanism within the thermostat sends a signal to the furnace to switch on when the temperature is lower than the

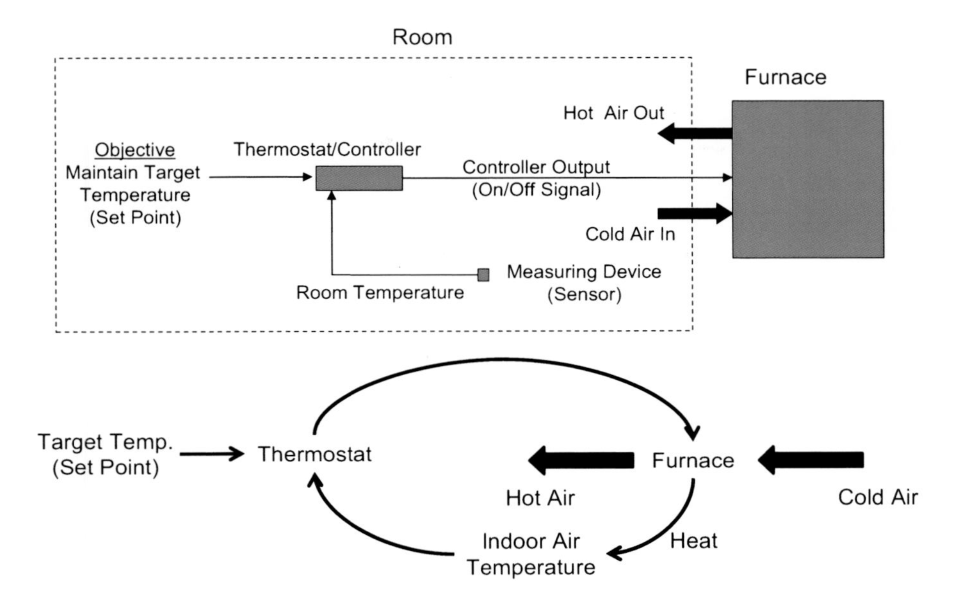

Fig. 1.2 A home heating system

desired value. The thermostat monitors the temperature continuously and when it reaches or exceeds the desired value, it sends another signal to switch off the furnace. Thus, the cycle goes on and the system acts like a loop. This is termed as a feedback control loop where the temperature in the room is the feedback signal.

1.3.2 A Bus Transportation System

A bus is a transporter, which is supposed to take its passengers from one place to another (Fig. 1.3). The bus is manned by a driver who controls its movements following the desired objectives, such as taking a preset route, stopping at stipulated locations for passengers, follow the traffic signals, and drive safely.

Here the driver is like the thermostat in a home heating system, but with a much more complicated set of tasks. The driver controls the speed and directions using his or her arms and legs to manipulate steering wheel, brake, and gas pedals. Here the driver's eyes and ears are the primary sensors, which are aided by speedometer, and sometimes by a GPS (Global Positioning System), to control the movements of the vehicle. Without these feedbacks, the driver would not be able to drive the bus properly and safely.

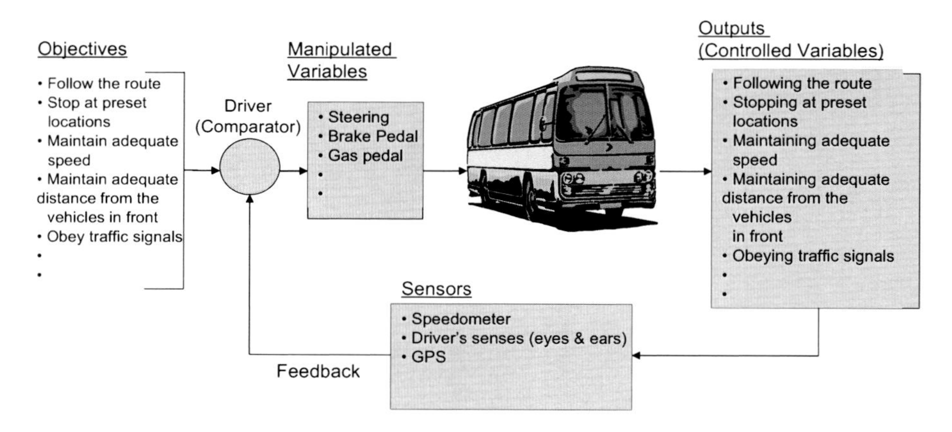

Fig. 1.3 A system for driving a public bus

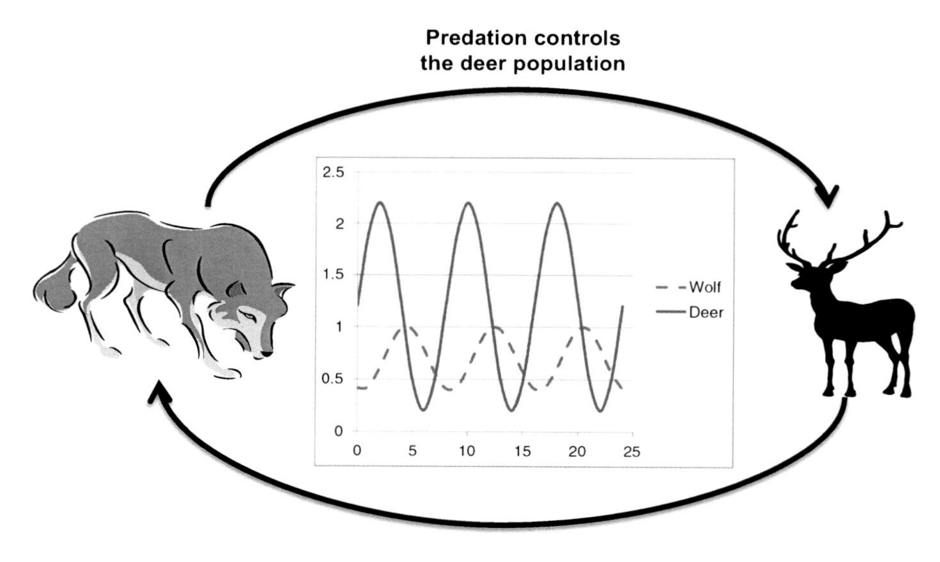

Fig. 1.4 A predator–prey relationship

1.3.3 A System of Predators and Prey

In many forests, wolves are the primary predators for deer. The deer population is controlled by wolves and the wolf population in turn is controlled by the availability of deer, which are their primary food source (Fig. 1.4).

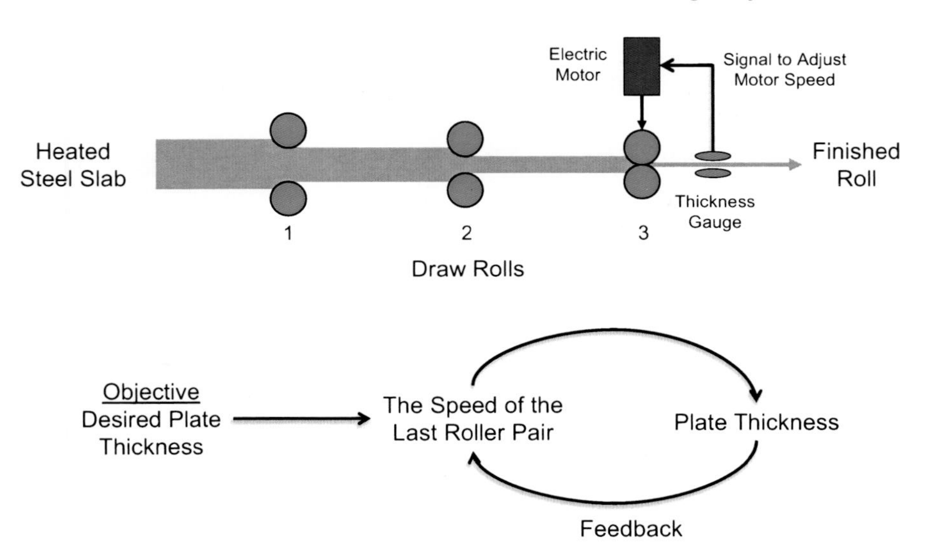

Fig. 1.5 Controlling the thickness of rolled steel sheets

Normally, there is equilibrium between the number of wolves (predators) and number of deer (prey). However, when there is a reduction in the number of predators the number of prey will increase. Predators, in turn, will increase as food is more readily available and thus more of their offspring will survive. Then, as the population of predators rises, more prey is killed and their numbers fall. Then, many of the predators may die because of the lack of readily available food. Thus, numbers of predators and prey may oscillate between two extremes. This is an example of oscillatory behavior that is found in many systems.

1.3.4 Controlling the Thickness of Rolled Steel

In a rolling mill, heated metal slabs are flattened into sheets, plates, or other shapes by passing them between heavy rollers. A typical steel rolling mill has a number of draw roll pairs through which heated steel slabs pass to produce sheet metals of desired thickness (Fig. 1.5). Thickness of a finished sheet depends on various factors, which include temperature of the heated steel slab, clearance between each pair of rollers, and the speed differences between pairs of rollers. For example, the difference of speed between an upstream pair of rollers and the higher speed downstream rollers causes a pull on the steel sheet tending to make it thinner. Additionally, any small difference in physical property of a steel slab affects the thickness of final product.

For years, mill managers and technical experts had been trying to improve the uniformity of rolled steel sheets, by trying to control all these factors more accurately. Some mill owners made enormous investments to do so yet their efforts largely failed. This is a typical case, where a desired result depends on a number of variables, which are mostly independent of each other. Trying to control one of these variables accurately is a formidable task, let alone all of them at the same time. By the time, correct adjustments were made to the clearance between the rollers and their speed, the temperature and other physical properties of the steel slabs being rolled could have changed requiring another set of readjustments to the rollers. Additionally, the wearing of a pair of rollers required frequent readjustments of their clearance. Therefore, a rolling mill had to be shutdown frequently to carry out all these adjustments, thus adversely affecting their outputs. In short, the technical experts and the management of rolling mills were in constant struggle to set these variables accurately in order to keep the thickness of rolled sheets uniform, which in many cases was a losing battle. Many rolled sheets went to scrap heap for recycling because their thicknesses were grossly out of the required specifications.

This problem, which had dogged the industry for decades, was solved comparatively easily when the measurement of the thickness of the rolled steel, as it came out from the last pair of rollers, was fed back to a mechanism that controlled the flow of electric power to the motor that powered those rollers. The continuous measurement of the thickness of the finished roll and using that measurement to control the speed of the last pair of rollers without any human intervention was a new idea. The rotational speed of the pair was increased automatically when the finished steel sheet was detected to be slightly thicker and conversely it was decreased when the sheet was found to be a little thinner than the required specification. The automatic regulation process was unconcerned about the reason for the deviation, which could be due to one or more of the factors stated earlier, but it did its job of correcting the deviation by just adjusting the speed of the last pair of rollers.

Once the problem of sheet thickness variability was worked out, the solution appeared to be obvious, which was quickly adopted by rolling mills around the world. It no longer mattered when other variables such as temperature of the steel slab or clearance between each pair of rollers changed (within reasonable limits). However, why did it take so long to come up with the solution? The technology for continuous and accurate measurement of the thickness of steel sheets and the ability to vary the speed of electric motors was there for quite some time but why did no one think earlier of using one to control the other? The plausible answer to this question is that engineers and technicians were focusing on controlling the variables that affected the thickness of rolled sheets individually, thus losing sight of the whole picture. When the system as a whole was taken into account the best place to intervene (leverage point) to get the desired result became obvious, which was the speed of the last pair of rollers. This was a triumph of integrated system thinking.

1.4 Similarities Between Disparate Systems

Some of these examples are gross simplifications of actual systems. There are additional factors that affect deer and wolf population than the example suggests. However, examining the main interactions allows us to understand basic system principles that may then be applied in more complex situations.

These examples are quite dissimilar in what they do and how they perform their tasks. However, viewing them as systems allows us to look at their similarities, such as the way their feedback, monitoring, and response functions work. That leads us to a better understanding of many other systems, such as a manufacturing process, political system, economy of a country, or the psychology of an individual or a group.

A system is more than the sum of its parts. These parts or subsystems are logically set to achieve one or more objectives. Thus, an automobile, which consists of an engine, four wheels, and a body, is designed for proper coordination of these components so that a driver can provide efficient and safe transportations of passengers and goods from one location to another. In a school, teachers, students, classrooms, and other amenities are coordinated to create an appropriate environment to impart knowledge along with emotional and cultural growth of our children.

A collection of parts or materials with no apparent objective is not a system. A pile of wood on a roadside is not a system or part of a system unless there is some objective, such as a farmer trying to sell the firewood to augment his income. Thus, whether an object is considered a system or part of a system depends on how it is perceived. Additionally, an object may belong to multiple systems. For example, the pile of firewood can be viewed as a part of the ecosystem that will increase the amount of carbon in the atmosphere or as a part of another system, that decreases the dependence on nonrenewable energy.

The dynamic nature of the systems and their components make system study so interesting. In many cases, it is easy to observe certain system inputs and their corresponding outcomes, but it may be quite difficult to fathom out their exact relationships. This is often due to incomplete knowledge of the systems or their behaviors. The relationships may not be direct or linear, for example, the relationship of students' achievements with the amount of money spent per pupil in a school district. Here the output, which is the students' achievements, has some relationship to the amount of money spent, but there are also many other factors involved that need to be considered.

System knowledge can help us better understand the social constructs we interact with, the machines we use, and the environment we live in. This allows us to improve or optimize them wherever possible or to be more adaptable to situations where we cannot change them.

1.5 Open and Closed Loop Thinking

We often do not realize the existence or the influence of feedback loops in our daily actions or in our interactions with the systems around us (Fig. 1.6). In some situations, one may take appropriate actions to solve a problem that he or she faces, and may not perceive the need to act any further when the problem is solved. This is an open loop situation, which is appropriate when the actions taken to solve the problem have negligible side effects. For example, if a farming community near a large river diverts some of its water for

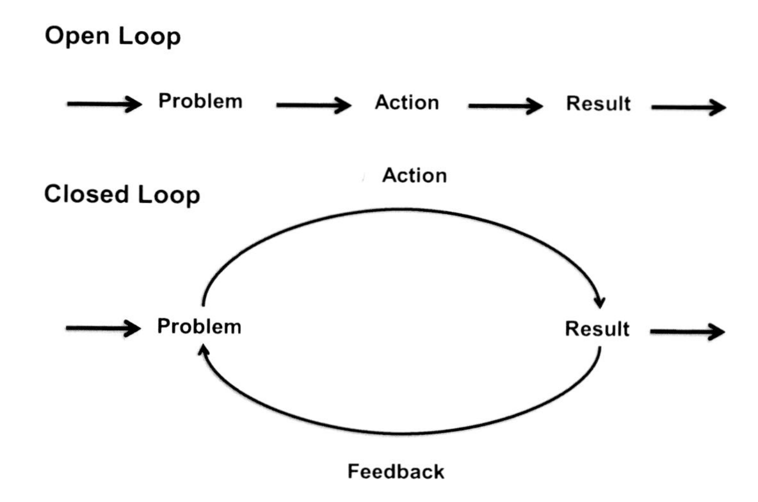

Fig. 1.6 Open and closed loop thinking

irrigation purposes and the amount of diverted water is quite small compared to the total volume of water carried by the river, then no further action is deemed necessary. However, when a large number of other communities start acting in a similar way there will be a significant reduction in the river water level. This will inevitably affect the irrigation systems of many of these communities, as well as cause biological and environmental problems that cannot be ignored. This then becomes a closed loop situation, where further actions are needed to take care of the consequences of the initial action.

Similar is the situation with our energy use. When only a very small percent of the population in this world owned automobiles and the known reserve of crude oil was plentiful, adding a few more automobiles had little effect on the price of gasoline or the quality of environment. However, with skyrocketing automobile ownership resulting in increased accumulation of green house gases, along with road congestion and traffic fatalities, the situation can no longer be ignored. This is very much a closed loop system, where actions are needed to respond to the consequences of increased car ownership.

In real life, closed loop is the order of the day. Very few situations may be characterized as truly open loop. Thus, one of the challenges in real life is one's ability to focus on the whole, rather than individual parts, with the realization that the complete nature of a system cannot be discerned without looking at all its component parts and their interactions.

1.6 System View

Modern scientists, technologists, and medical personnel generally follow an analytical process, where they focus on smaller and smaller areas of their fields of specialization. They generally take the path of reductionism, which states or assumes that a complex system can be understood and explained by reducing it to its fundamental parts. Scientists focusing on smaller and smaller elementary particles, technologists producing discrete components of a device, and doctors specializing in diseases that affect particular organs in our body are all taking the reductionist approach. By doing so they have improved our basic understanding of organs, created highly sophisticated tools, and invented new medicines, which offer enormous benefits to the society. However, this increased compartmentalization and narrow focus has led to the neglect of the big picture, which includes our industrial, social, economic, and natural systems; their interrelationships and how they are affecting us today; and how they will affect our children and grandchildren tomorrow.

Narrow thinking is not limited to scientists and technologists. It pervades every level of society, from politicians to army generals and bankers; and it is even prevalent among philanthropists. As our societies and their workings become more complex, we are creating numerous silos or compartments, where each specialist focuses his or her own talents to solve the part of a problem that is best understood, with little communication with others who may be working in the related fields. At the same time, our societies are getting increasingly interconnected, by the ease of travel and the increased supply of goods and services around the world. Therefore, the consequences of actions by one individual or a small group have ripple effects on the society or even around the world.

There are many examples of local problems that have had much wider implications, such as the SARS epidemic, which was due to overcrowding of human beings and their livestock, the pollution by a small group of industries in newly emerging countries, or the US army toppling the regime in Iraq without making adequate plans for social reconstruction. The worldwide financial debacle in the early part of this century also has its roots in such silo thinking, where mortgage lenders and bankers did not fully understand the economic implications of their actions or did not care about their longer-term implications.

Compartmentalization of thinking has resulted in numerous actions with unintended consequences. One well-known example of inadvertent harmful consequences of international philanthropy is in Bangladesh, where deep wells were drilled in the 1970s at the behest of UNICEF to provide clean drinking water. It was widely believed that water from shallow wells is more contaminated than from those that are much deeper. This turned out to be a narrow view as no effort was made to test the deeper aquifers for contaminants. It was not until the early 1990s that people in Bangladesh became concerned about increased frequency of skin lesions and other health problems affecting people in many towns and villages. That led to the testing of water from these deep wells, which showed contamination of heavy metals and high levels of arsenic. Authorities in Bangladesh and neighboring areas of India are still struggling to develop adequate water filtration systems to remove arsenic from thousands of those deep wells and to treat the ensuing cancer in those who drank water from these wells.

Another example of unintended consequences is the use of methyl tertiary butyl ether (MTBE) as an additive to gasoline in the USA. MTBE is a volatile, organic chemical, which promotes more complete burning of gasoline, thereby reducing carbon monoxide and ozone levels in exhaust gases. It was added to gasoline in certain areas of the country that did not meet air quality standards set by the Federal Clean Air Act. While MTBE has succeeded in reducing air pollution, leaks of the chemical from underground gasoline storage tanks have polluted nearby water sources in many places. Because the US Environmental Protection Agency (EPA) considers MTBE a possible human carcinogen, a number of states have now banned its use. However, the problem of groundwater pollution persists and it will take a lot of money and effort for its cleanup.

System thinking, in contrast to traditional reductionist approach, takes a wider view in solving a problem and is thus becoming increasingly useful in many areas, such as large scale engineering designs, medicine development, environmental and ecological problems, social and political problems, economics, human resources, and educational systems, which we will explore in the following chapters.

1.7 System Types

There are many different types of systems in this world. Commonalities can be found among all of them in regard to the way they behave. However, they may be classified according to the way they are built and how their control mechanisms function (Fig. 1.7).

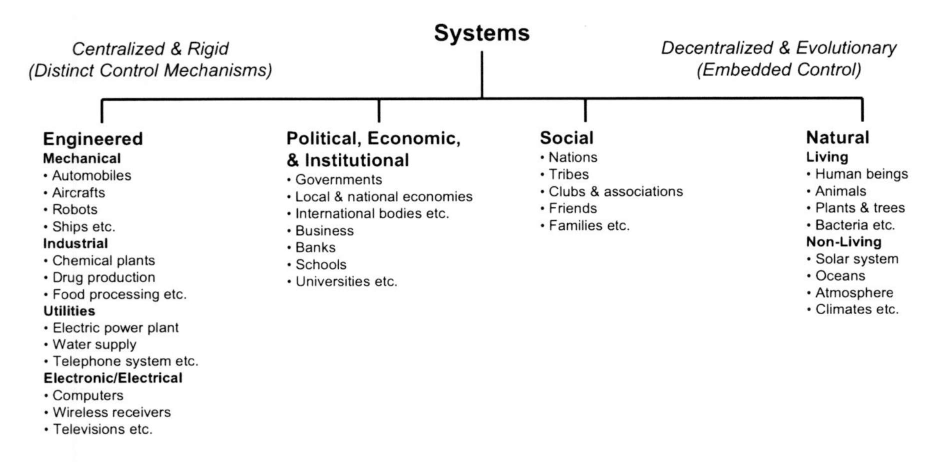

Fig. 1.7 The system type continuum

1.7.1 Engineered Systems

There are industrial systems, such as chemical plants, oil refineries, plants for food and drug manufacturing, manufacturing of automobiles, or textiles. There are electromechanical systems, such as ships, airplanes, automobiles, and rockets, and then there are utilities, such as water and electric supply systems. All of these manmade systems may be grouped together as "Engineered Systems." For their proper working, they all need control mechanisms that are generally distinct from their functional parts. These controllers may be mechanical, electronic, electromechanical, or manual/human. An engine provides necessary power to an automobile but that is controlled by accelerator (gas pedal), clutch, and brakes, which are separate entities. Chemical plants and oil refineries require complex control systems along with sensors and electromechanical devices (actuators) to control the flow of material and energy along with human supervision for their proper working. Airplanes and ships have sophisticated control devices onboard along with trained human pilots to navigate them.

It is important to recognize that the controllers that control the behavior of engineered systems are usually distinct entities (subsystems). However, as we will discuss in the following sections, that is not usually the case for natural living or nonliving systems or for many political, economic, or social systems.

1.7.2 Natural Systems

In contrast to engineered systems, there are natural systems, which include living systems like human beings, animals, plants, and bacteria and nonliving systems, such as solar system, rivers, oceans, earth's atmosphere, and climate. The control mechanisms for these natural systems follow laws of nature and physics, which are distributed and embedded within these systems. A tree grows from a seed in the right weather and grows according to its genetic code, subject to the availability of proper soil, water, and other nutrients. A maple tree sheds all its leaves at the start of winter and starts growing them again in spring while producing sap to help the growth. That requires very sophisticated control mechanisms. However, the tree's control mechanism is entrenched within its system. In the case of a living being of a higher order, such as a human being or an animal, one could argue that the head controls its body, creating separation between control and actions. However, the two functions are so intimately intertwined, that it is difficult to separate them.

The behavior of a nonliving system, such as earth's atmosphere or a river is based largely on physical laws but is also affected by human activities.

Deforestation and burning of fossil fuels increase CO_2 concentrations and various industrial activities increase other forms of pollution that are leading to climate changes. Thus, the control mechanisms for natural systems follow physical laws but are largely decentralized.

1.7.3 Social, Political, Economic, and Other Types of Systems

In between these two extremes of engineered and natural systems, there is a whole host of others. These include social systems like nations, tribes, friends, and families as well as political and economic systems, such as governments and economies, local and national. Then there are institutions, such as businesses, charitable foundations, schools, colleges, and universities. Where they lie within this continuum of system types depends largely on their structures. A highly decentralized system such as a cooperative enterprise may be more like a biological or natural system, as would be a fully functional democracy. However, a business with a rigid top down architecture or a centralized dictatorship would be more akin to a common engineered system. Interestingly, the Internet is an anomaly; it is an engineered system but highly decentralized with no fixed hierarchy.

1.8 Information Is Fundamental to Systems' Behavior

A close examination of any system, whether it is engineered, social, or natural, shows that information exchange between its components or subsystems plays a crucial role in its workings. In the case of a central heating system, the thermostat gets information about the indoor air temperature, which it compares with the desired temperature, to generate appropriate on/off signals for the furnace. In a system of predators and prey, the number of living animals of each type, at a given time, is the information that largely determines their composition in the future.

Materials and energy do flow between many interacting systems and subsystems, such as flow of gasoline from a tank to the ignition chamber, or transfer of energy from the engine to the wheels of an automobile. Similarly, there are flows of electrical or optical energies in a telephone system when two persons communicate. However, system study has more to do with the flow of information (signals) than with the transfer of material or energy.

Thus, a system or a subsystem's actions at any given time are largely based on its own state and the information it receives from other systems or subsystems that interact with it. A system's own state, however, may have a lot to do with its past experience. For example, the rate of flow of water in a stream may vary with the amount of rainwater presently pouring into its catchment area, but equally important may be the total rainfall in the last rainy season. The behavior of an insect is largely dictated by its instincts and the stimuli (signal) it gets at a given time, while a dog may react to a given situation based on his past experience and training. The behavior of higher animals, like human beings, not only depends on their current stimuli but also on their genetic makeup, past experiences, childhood learning, and societal norms. Simple mechanical systems react in predictable fashion to a fixed set of stimuli, whereas an intelligent machine can learn and behave differently with increased experience.

This information or signal may take various forms. Communication between humans is predominantly verbal, written, or visual. Animals may use sound, scent, and vision to communicate. Various forms of physical energy are used for communications in engineered systems, which include electrical, thermal, and chemical. Living plants may use various visual and chemical signals to communicate with other plants or for attracting pollinating insects and birds, such as the color and scent of their flowers.

In an engineered system, the exchange of information or signal is usually distinct from the transfer of materials or energy. In the case of a central heating system, the furnace consumes fuel to heat the air in a building, but the information flow is distinct from the flow of fuel or the flow of heated air or water. However, in many biological or natural systems such distinctions are not obvious or are nonexistent. In the case of a system of predators and prey, the wolf and deer populations at a given time may be considered as both material stock as well as signals for the behavior of the system. In the case of the carbon cycle, the amount of global warming depends on the actual concentration of CO_2 in the earth's atmosphere at a given time.

Thus, in order to appreciate how and why a system behaves the way it does, it is necessary to analyze the information flow within the system along with those that are between other interacting systems.

1.9 The Importance of System View

Viewing the world through the lens of a system has many benefits, which are summarized here.

A System Is More Than Its Parts A system is a collection of parts that interact to function as a whole. This concept, that the whole is greater than the sum of its parts, has been known from the time of Aristotle.

The total value of a system often exceeds the combined worth of individual components of that system. For example, a house is worth more than the value of the individual pieces of wood, sheetrock, drywall, wiring, and other components that were used to build it. A pack of wild dogs is more effective in hunting than the sum of the individual hunting capabilities of each dog in that pack.

Dealing with Complexities A complex and dynamic system interacts with many other systems and is composed of many subsystems that interact with each other. Understanding basic system principles allows us to understand the present and future behavior of a complex system better than would otherwise be possible. For example, the phenomenon of global warming and weather changes may be better understood by looking at the carbon cycle as a system with everything that interacts with it. Similarly, the understanding of economic booms and busts may be better appreciated by the understanding of business cycles.

Interactions with Engineered Systems Engineered systems are all around us and we interact with them all the time. They range from automobiles to elevators and home heating to water supply. Sometimes we are awed by the workings of airplanes, robots, and rockets. How an airplane lands perfectly (most of the time) in near zero visibility condition or how drones (unmanned aircrafts) can perform reconnaissance and pinpoint enemy targets. In addition, how robots can perform boring repetitive tasks or those that are highly dangerous to human. System perspective lets us better understand them and use them optimally.

Ways of Dealing with Systems Over Which We Have Little Control We cannot change or modify most of the systems we deal with in our everyday life, such as traffic jams, gasoline price, or bureaucracy. However, system knowledge gives us the tools to better understand the problems and allow us to formulate improved strategies for overcoming them.

Improve Systems Over Which We Have Some Control System knowledge allows us to improve those things over which we may have some control. It allows us to identify the leverage points in a system, where making small changes may lead to much greater beneficial effects. For example, delays or bottlenecks within a system play a major role in its inefficiencies. Reducing or eliminating them wherever possible can produce significant improvements to their workings. Not all bottlenecks or delays are avoidable, but system knowledge can help us decide which of them can be eliminated or reduced, and which one should be tackled first. Similarly, closer coupling and faster communication between systems and subsystems may lead to significant improvements in their workings.

Create Sustainable Systems Proper understanding of basic physical laws and our ecological systems helps us in taking appropriate actions to safeguard our own future and those of our children. System knowledge allows us to understand the full lifetime cost of a manufactured product, which includes material and energy usage not only for its manufacture but also during its entire lifetime of use, and also for its disposal at the end of its useful life. Thus, a product that uses less energy may not improve sustainability if it causes significant environmental problem for its disposal.

The ability to control mechanical and electromechanical systems properly and precisely plays an essential part in an industrial society. Many of them are complex systems with many interacting parts that need to work properly to offer quality, efficiency, and safety. Greater focus on sustainability is leading us to efficient use of energy and raw materials, reducing wastage, and reducing production of harmful by-products. Improving the controls of these systems will play a very important role in achieving those goals.

How to Avoid or Manage Unintended Consequences When we face a problem we take necessary actions for its solution but often we do not understand or ignore the secondary or longer-term effects of those actions. Thus, when the demand for fish goes up in a region, the fishermen start fishing more and get prosperous, but that leads to the depletion of fishing stock, affecting the livelihood of those very fishermen a few years later. Broad system thinking can avoid or minimize these unintended problems by looking at all the major implications of an action in the shorter and longer time horizons.

1.10 The Way Forward

In the following chapters, we will look at various systems in detail and their fundamental behavior patterns. We will examine the central role of feedback in a system and consider how inadequate, erroneous, or delayed feedbacks can degrade them. We will also discuss why systems may oscillate and become unstable or produce unacceptably sluggish responses.

We will model some of these systems and look at ways to optimize their behaviors to make them more efficient. We will be looking at centralized systems first because they are easier to discern. Then we will consider decentralized systems and their relative advantages over the centralized ones.

We will touch on discrete events and procedures and their effects on system behaviors. We will consider various unintended consequences that can happen when we try to change or modify a system along with ways to overcome them. In the final chapter, we will discuss how a system-oriented person would look at various problems and find ways to solve them.

2

Engineered Systems

Abstract Engineered systems are designed and constructed by human beings. They include a vast array of mechanical, electromechanical, electronic, and hydraulic devices, such as steam engines, automobiles, stereo-amplifiers, computers, and wind turbines. They also include much larger entities, for example, ships, airplanes, chemical manufacturing plants, and oil refineries, and even larger ones, such as telecommunication networks and power grids. An engineered system may include human beings, when they are closely associated with its working, for example, an automobile with its driver, an airplane with its pilots, and a chemical plant with its operating personnel. The early applications of system science have mostly been in the domain of Engineered Systems. James Watt was a forerunner who developed feedback mechanism for controlling the speed of steam engines. This chapter provides an overview of engineered systems as an introduction to broader system thinking.

The chapter begins with a brief history of engineered systems that came into being at the start of the Industrial Revolution. Then there is discussion of how Watt Governors controlled steam engines and why that played a significant role in the early days of engineered systems. That is followed by discussion of loop diagram and how it helps in depicting the information flow in an engineered system. The chapter delves into the various parts of a loop, which include sensors, actuators, and controllers. Finally, the importance of feedback is discussed followed by the possible negative effects of delays.

A. Ghosh, *Dynamic Systems for Everyone*,
DOI 10.1007/978-3-319-43943-3_2

2.1 History of Engineered Systems

The Industrial Revolution, which started in Great Britain, during the eighteenth century, totally changed the ways of producing goods and services for human societies. It also changed the society from being mainly agricultural to one in which industry and manufacturing became paramount. Soon after its adoption in England, other countries such as Germany, the USA, and France joined in this revolution. Then, Japan and Korea led the industrialization in Asia, which is now being followed by other major countries, such as China and India.

Industrial Revolution led to the design of engineered systems on a large scale. These ranged from manually controlled to partially and fully automated systems. Before the Industrial Revolution, human societies were largely agricultural with artisan-based small-scale industries, also known as cottage industries. In the cottage industry era, which spanned for a very large part of humankind's history, there was hardly any need for mechanical feedback systems, as human beings controlled the production processes. That changed with the Industrial Revolution, when machines replaced a significant amount of human activity. The machines not only replaced the muscle power of humans and animals, such as horse and bullock but also started to reduce the need for human intervention for their operation. This drive toward automation and the need to operate with greater efficiency led to the systematic study of their behaviors.

The early mechanical systems were water wheels and windmills, but the big jump in mechanization occurred when steam engines came into use. At that time coal, which was abundant in England, was the fuel of choice for generating steam. The first use of stationary steam engines was for pumping water from the coalmines in England where digging deeper mines was increasing the water logging problem. Each of these engines replaced hundreds of horses that were formerly used to raise the water. These engines were large and stationary, which were difficult to operate at a constant speed. When a boiler for generating steam became too hot it produced excessive amounts of steam making the engine run too fast. As controlling the coal fire was not easy, often cold water had to be injected into a boiler to control its rate of steam production. This was wasteful, which led James Watt to design a system with a fly ball governor for regulating the flow of steam to an engine and thus control its speed. This device, commonly known as the Watt Governor, holds a central role in the paradigm shifts that occurred in various manufacturing processes during the Industrial Revolution.

James Watt

The person, who is remembered above all for improving the efficiency and controllability of steam engines at the start of the industrial revolution, was James Watt, a Scottish engineer and an instrument maker. He was instrumental in ushering the age of mechanical power. His inventions, which successfully combined science and technology, contributed significantly to the industrial revolution.

James Watt was born in 1736, in Greenock, Scotland. He started working from the age of 19 as an instrument maker and soon became interested in improving the performance of steam engines. Watt designed a separate condensing chamber for the steam engine that prevented enormous losses of steam in the cylinder and enhanced the vacuum conditions. Watt's first patent, in 1769, covered this device and other improvements on the engine, which included steam-jacketing, oil lubrication, and insulation of the cylinder in order to maintain the high temperatures necessary for maximum efficiency. From a system perspective, Watt's most important invention was the fly-ball governor for controlling the speed of steam engines, which is generally known as the Watt Governor. Watt made several inventions in various other fields, such as in civil engineering and telescopy. Watt died in Heathfield, England, in 1819. He changed the course of the industrial revolution and was the most honored engineer of his time.

2.2 The Watt Governor

The Watt Governor consists of two iron balls located at the end of two pins that are connected to a spindle, which is attached to the flywheel of a steam engine (Fig. 2.1). There are two rings/sleeves that connect to the spindle, where the upper sleeve is rigidly connected and the lower sleeve slides up or down. As the velocity of the wheel increases, the centrifugal force causes the two balls to separate further and further, which makes the lower sleeve to move up causing reduction of the steam flow to the engine and lower its rotational speed. Conversely, when the engine slows, the two balls get nearer to each other resulting in increased steam flow and its rotational speed. Thus, the rotational speed of the steam engine is kept reasonably steady around the desired set point. The vertical position of the upper ring/sleeve is adjusted to vary the set point of the rotational velocity.

The Watt Governor gave factories and mills reliable and cost-effective ways to regulate the speed of various types of machines, from water wheels to steam engines. Later on, the principles of the Watt Governor were applied to control various other machineries, such as internal combustion engines and turbines.

The Watt Governor is notable as it is one of the first engineered systems using feedback principle. This was the first time that the use of mechanical

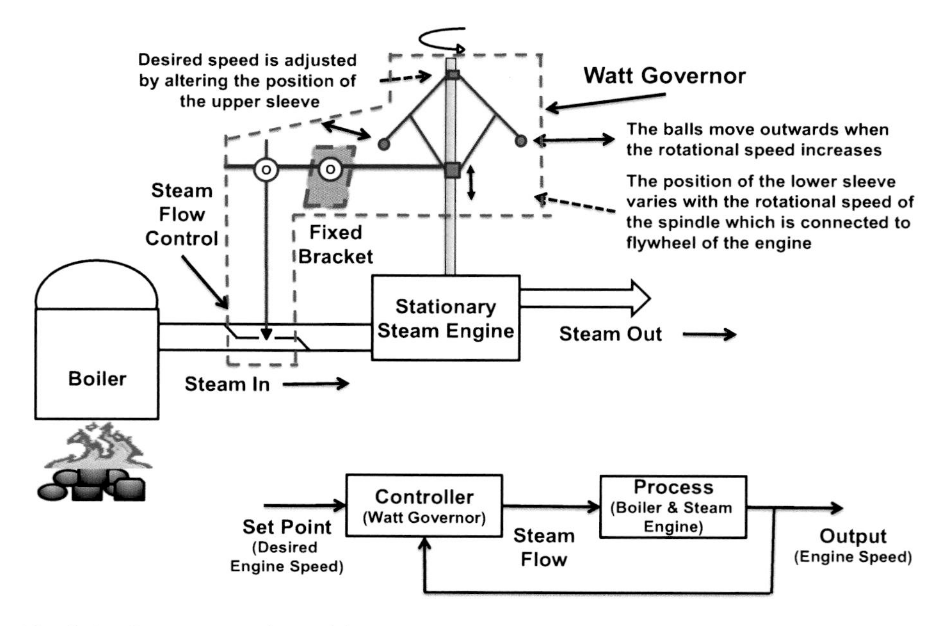

Fig. 2.1 A steam engine with a Watt governor

technology was extended from power to regulation. It improved the reliability of steam engines significantly by automating the manual functions of an operator. Without the Watt Governor the Industrial Revolution could not have progressed, as steam engines lacking automatic control would have remained very inefficient, requiring much more manual labor and energy to operate them.

The Watt Governor is of particular interest to us because it embodies the principle of feedback, linking output to input, which is the basic concept of automation in almost all systems. James Clerk Maxwell, the famous Scottish physicist and mathematician, made extensive analysis of its principles a century later.

2.3 Depiction of Information Flow (Control Loop Diagram)

In studying the behaviors of a system, we focus on the actions and interactions of its components or subsystems. These actions and interactions are largely the result of information or signals that flow within and between them. Textual narratives, while often sufficient to describe linear events, fall short of

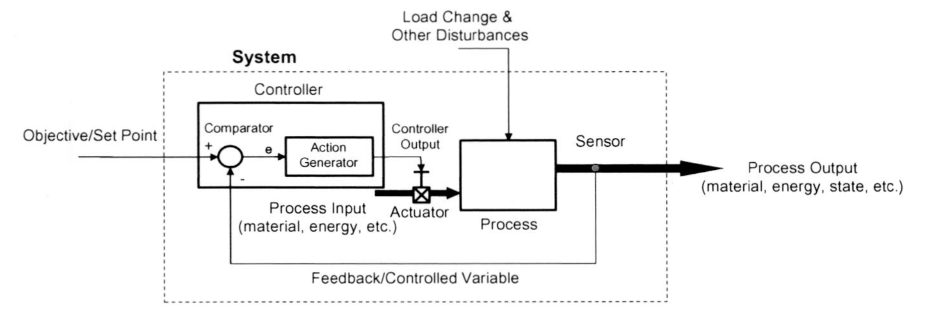

Fig. 2.2 Loop diagram of a system

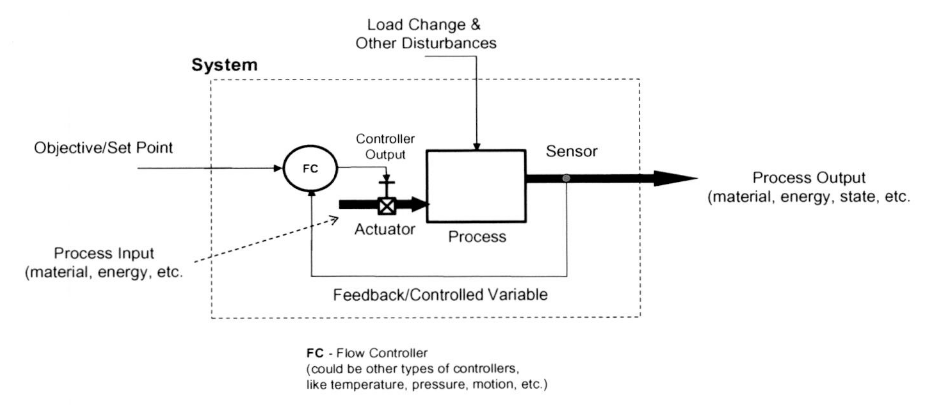

Fig. 2.3 Loop diagram of a system (alternate format)

depicting these information flows adequately. This has led to graphical representation of systems, which are often supplemented by texts.

Engineered systems are commonly represented by loop diagrams as shown for the home heating system in Chap. 1 and for the Watt Governor in this chapter. In a loop diagram, rectangular and circular blocks are used to represent functions or subsystems, which are connected by arrows, which depict the signal or information flows and their directions (Figs. 2.2 and 2.3). An arrow is usually annotated to describe the type of information that it is transmitting. Sometimes, a positive or a negative sign is shown at the head of an arrow when two or more of them are compared. Material or energy flows are sometimes shown in block diagrams, which are distinguished from information flow by thicker sets of arrows. Physical objects such as valves, pumps, or process vessels are often illustrated by blocks to make them easier to understand. Loop diagram is a preferred way to depict engineered systems and is widely used in technical literature.

2.4 Going Round a Loop

In the following sections, we will be looking at the different parts of a typical control loop for an engineered system. The understanding of which will give us more insight about the workings of all types of systems.

A simple engineered system consists of a process and a controller. A process may be any subsystem that does something that is useful or interesting, which is regulated by a controller. The home heating furnace, steam engine, automobile, and chemical reactor are examples of such processes. A simple controller controls one of the variables of the process to produce the desired result, which could be the room temperature for a home heating system, the desired speed in case of a steam engine, or desired product composition and quality in the case of a chemical reactor. The controller may be an electrical or mechanical device for a home heating furnace or a steam engine, but it also can be a human driver in the case of an automobile, or an airplane.

The block diagram of a simple control loop is suitable for depicting situations where one subsystem is controlling another, such as between an automobile and its driver, a teacher teaching a pupil, or a mother tending her baby. Sometimes the controller and the process are interchangeable entities (subsystems), depending on their roles and the way we perceive them. For example, in an interaction between a mother and her baby the mother is generally in control, while the baby is the process. However, the baby may be considered as a controller whom the mother feeds whenever the baby screams.

There is another type of relationship, which is peer-to-peer, such as the interactions between two friends, between a married couple, or between a parent and a grown-up child, where each actor may be trying to influence the other (Fig. 2.4). Then, there are two set points or objectives, which may or may not coincide, leading to a more complex situation.

2.4.1 Objective or Set Point

Every engineered system has one or more objectives. A home heating system's function, for example, is to maintain the temperature in a house at a set value (set point). The objective of a Watt Governor is to run a steam engine at a set speed. In a manufacturing plant, the objective may be to maintain the quality of its products. For an enterprise, it may be to optimize profit.

In a primary school, a teacher's main objectives are to impart literacy, numeracy, creativity, and communication skills along with the enjoyment for learning.

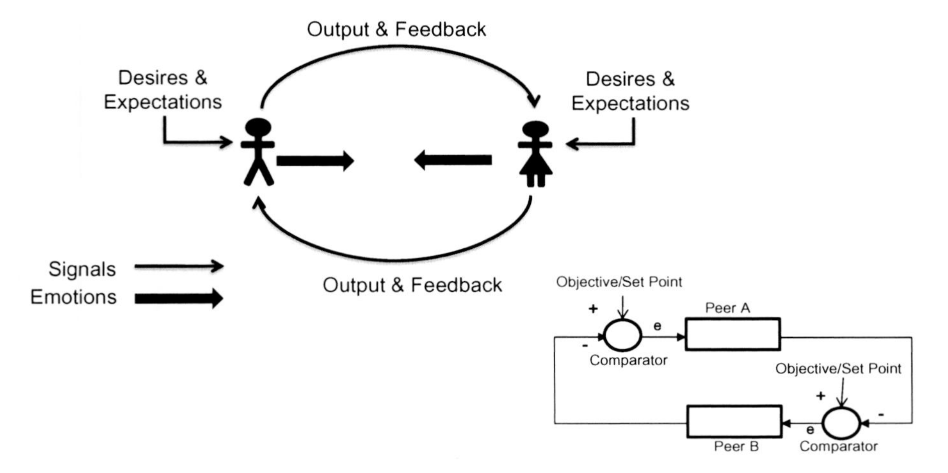

Fig. 2.4 Peer-to-peer interactions

Whereas, in a high school the objectives may be different, such as imparting in-depth knowledge in chosen subjects, making the student a good corporate citizen, and maximizing a student's prospect of gaining admission to a prestigious university. A politician when elected to a high office may have many different objectives. Many of them may be the promises made during the election campaign but his or her primary objective may be to increase the prospect of getting reelected. Thus, in a typical system, there may be multiple objectives. However, to simplify the matter, we will initially consider only one objective at a time and leave the consideration of multiple objectives for a later chapter.

2.4.2 Controller

A controller consists of a comparator and an action generator. Some people depict them separately, however in real life they are mostly together, such as the controller of a home heating system, the Watt Governor for a steam engine, a driver driving a motor vehicle, or a teacher teaching a pupil.

Comparator

The task of a comparator is relatively simple; it is to compare the objective or the set point with the controlled variable (feedback). The result of the comparison is a positive or negative value, which represents the amount by which the actual output of the system varies from the target or the set point. The difference between the desired and the actual values is called error (*e*).

Though comparator has a distinct function, in most cases it is physically a part of the controller. Thus, the driver of an automobile functions both as a comparator, comparing the speedometer reading with the desired speed, and as a controller adjusting the position of the accelerator.

Action Generator

An action generator generates the controller output, which is based on the error signal from the comparator. In the case of a home heating system, the thermostat (controller) generates either an on or an off signal depending on the current room temperature in relation to the set point. Appropriately, it is called on/off control and such a controller as on/off controller. On/off control works well in maintaining the temperature in a room as human beings can generally tolerate a small variation in the ambient temperature without much discomfort (Fig. 2.5). On/off control will, however, be undesirable in other applications, such as for driving a vehicle, where smooth ride is an essential element for comfort. It will also be unacceptable for controlling temperature in an industrial application, where the quality of a product is closely dependent on the temperature during the chemical reaction.

In industrial applications, proportional control action is most common, where controller output is proportional to the error. That is, further the process output is from the set point the harder is the push to get there. For example, we increase pressure on the gas pedal when a car is going slower than the desired speed. The amount of increase in pressure we put is proportional to how much we want the speed to increase. Similarly, the pressure we put on

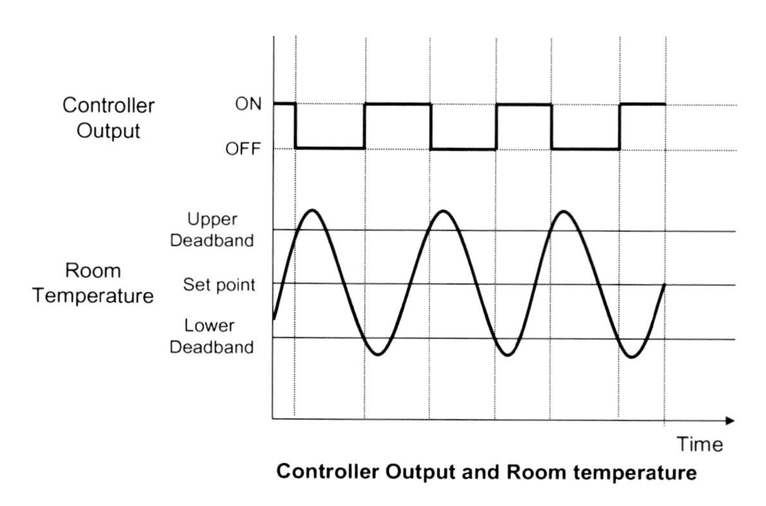

Fig. 2.5 Behavioral response of a central heating system

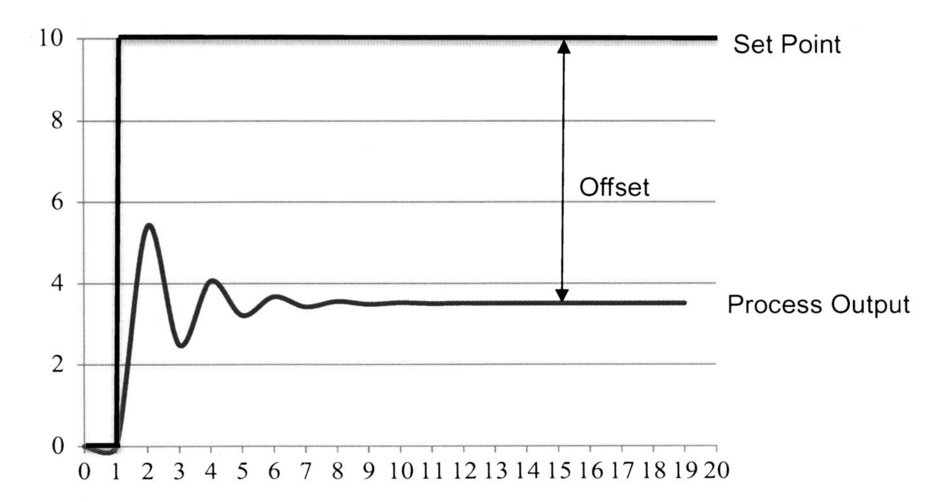

Fig. 2.6 Proportional control with offset

the brake pedal is proportional to how fast we want the car to slow down. Proportional action is also used in social settings. Our criminal justice systems are largely based on proportional control. When we say, "the punishment should fit the crime" we are talking about proportionality.

The main drawback of pure proportional control action is that it produces output only when there is error. That leads to a situation where the output never reaches the set point. Once the process output gets closer to the set point, the error gets smaller and the output of the controller gets smaller too, which may not be enough to get all the way to the set point or the target value. This difference between set point and controlled variable is called offset (Fig. 2.6). The offset can be reduced by increasing the proportional gain but that can lead to oscillation, which will be discussed in a later chapter.

Offset is generally corrected by adding another function called integral control action, whose output is based on the sum of the past errors. Integral control closes the gap by adding up (integrating) the error and the length of time the error has been there, thus eliminating the offset altogether. While driving a car we do that unconsciously by increasing the pressure on the gas pedal when it runs at a slower than the expected speed and then easing off when the expected speed is reached. Integral action is usually active in criminal justice systems where repeated offenders are given stiffer sentences than those who are first time delinquents. In industrial applications, integral control is commonly used along with proportional control because integral control is typically slower to respond on its own.

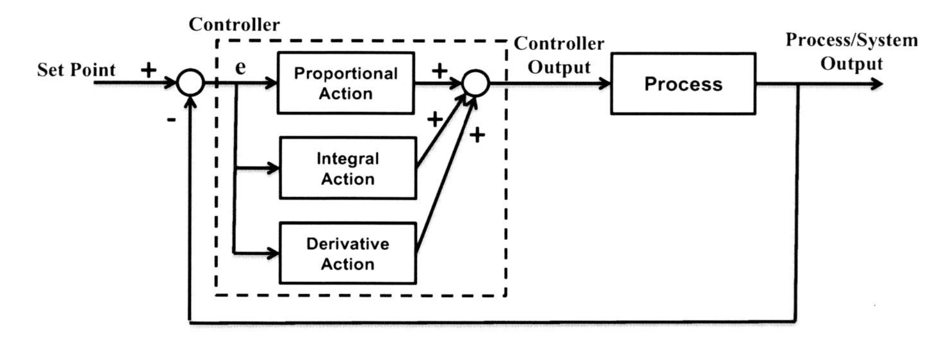

Fig. 2.7 PID controller actions

There is still another common type of control action called derivative control, where the output is based on the rate of change of an error. In other words, the magnitude of its output is greater when there is a faster change of the error signal. Thus, if the error increases suddenly, derivative action will act immediately, while it may take some time for proportional and integral actions to take effect. For example, we press the brake pedal when we see a sudden slowing of the car in front. The faster the car in front slows more pressure we apply to the brake pedal.

Proportional control is most common in engineered systems and integral control is used in conjunction with proportional control to eliminate the offset, where needed. Derivative control is used in conjunction with the other control actions where there is the need for rapid actions for sudden disturbances. A controller with all the three control actions is called a proportional-integral-derivative (PID) controller (Fig. 2.7). (For more details on PID control, see Appendix I.)

2.4.3 Actuator

An actuator converts the signal from a controller to something tangible such as the pressure on accelerator, altering the flow of gasoline to a car engine. A water tap is an actuator that alters the flow of water depending on its rotational position. An on/off electric switch is an actuator that depending on its position allows or stops electric energy to flow to a lamp, an appliance, or a motor.

Many different actuators are used in industry, which include valves for controlling flow or pressure of fluids, switches for stopping or allowing the flow of energy, and servomotors for accurate positioning of objects. Human beings use arms and legs as actuators and so do many animals.

2.4.4 Process

As stated before, a process is a subsystem that does something useful or interesting. A home heating furnace, a student in a learning mode, an automobile travelling from one place to another, and a steel rolling mill are examples of processes. Their dynamics vary with their inherent nature and their setup. What we mean by dynamics is the way the output of the process changes with change in the set point, input, or its load. For the home heating furnace, the set point is the desired temperature, the input is the room temperature, and the load is the total heat required to keep the room at the right temperature. The load may change when the outside temperature changes or if someone opens or closes a door or a window causing sudden change in the room temperature.

The dynamic response due to any of these changes may be instantaneous but more commonly there would be a lag, which may be in seconds or minutes for a home heating system or a manufacturing plant, whereas it could be months or years for an economic or a social system.

2.4.5 Process Output and Sensor

In the case of a home heating system, the process output is the hot air, for a steel rolling mill, it is the rolled steel plates, and for a chemical plant, it is the chemical produced.

Sensors are essentially converters. A sensor reads the process output and converts it to a signal that is easy to read, understand, or interpret. The speedometer in a car is a good example; it converts the circular motion of a wheel to the speed of the vehicle in a way that makes it easy for the driver to understand. Dogs' powerful olfactory (smell) sensors are useful to us in many ways, such as for finding a lost person or detecting contraband materials in a package. We use all our five senses to our fullest capacity in getting along in our everyday life and when one is impaired, we unconsciously augment others to compensate for the loss.

The human brain is very efficient at interpreting these senses but our sensors are not always accurate enough or repeatable. This is because we humans are complex systems, which try to compensate for a new signal based on our previous experience. This was illustrated by a simple experiment that I did in my middle school years, which may be repeated easily. It requires three coffee mugs, each about half filled with water, one ice cold, another lukewarm, and the third hot (not too hot to burn your fingers). You then dip one finger of one hand in the ice-cold water and one finger of the other hand in the hot water. After keeping them there for about a minute, put both the fingers in the lukewarm water. You will find that one finger feels hot and the other cold. This shows that our senses are very strongly influenced by our previous experiences.

There are numerous types of sensors used in industry. They include sensors for temperature, pressure, speed, flow rate, level, composition of a fluid, and the position of objects. Many of these sensors operate continuously and give results instantly (in real time). Others, such as gas chromatographs that accurately measure chemical compositions of gases, work in discontinuous fashion. Sensor accuracy is of great importance in the running of an industrial system because it produces the feedback, which makes the controller to act properly.

In our society, we use various sensors to get appropriate feedback. Periodic tests taken by pupils are sensors for teachers, as are opinion polls for politicians, and the census for long-term planners. Sometimes sensors and feedbacks seem to be intertwined, such as an MRI test done on a patient under a doctor's instruction. However, for purists the MRI machine is the sensor while the test result is the feedback to the doctor.

Many of these sensors, in the social context, work intermittently thus generating new results after certain time interval that, however, is changing with the progress in communications. Email and instant messaging are altering the landscape of opinion polls conducted by radio and television stations, which are having ramifications in our present societies.

2.4.6 Feedback

Feedback is a fundamental concept in system science. It is the information about a variable that the system is trying to control. Thus, when driving a car the driver controls its speed, while the speedometer provides the feedback. Similarly, a cruise controller keeps the speed of a car at or near a set value by adjusting the fuel input to the engine based on its speed.

Feedback is everywhere and we use it, mostly unconsciously, in our actions all the time. We use feedback to lift a cup of coffee from the table. While an arm moves toward the cup, the relative positions of the arm and the cup are fed back to our brain through our eyes. The brain commands the arm to move in the right direction at a comfortable speed so as not to overshoot the target and commands to grab the cup when the target is reached. A visually impaired person uses tactile senses as feedback to carry out the same function.

2.5 Right Feedback to the Right Person at the Right Time

The way a manufacturing company measures its profitability is through its cost accounting system. Traditionally, a cost accounting system consisted of bookkeepers and accountants manually performing the necessary calculations.

The main objective of such a system is to report the financial performance to the management and the investors, where the quantity of a manufactured product, for a given time period, is entered along with the cost to make those products. Cost accountants then calculate the total manufacturing cost to arrive at the unit cost of a product (Fig. 2.8). Today, these calculations are performed by computers but in most cases, the basic structure remains the same.

This method of cost accounting, however, has little relevance to the personnel that are responsible for the day-to-day manufacturing operations. When a system can do these calculations in real time and make them available immediately to the operating personnel then the data can be used to improve the operations in ways that can reduce energy needs, minimize raw material wastage, and improve quality, thus minimizing the cost of production. This real-time feedback of cost accounting calculations is called dynamic performance measures or DPM (Fig. 2.9). It should be mentioned here that not all cost accounting calculations results are needed for the operating staff, as that may be too much data, but only those that that are keys to measuring the performance of the system in the real time.

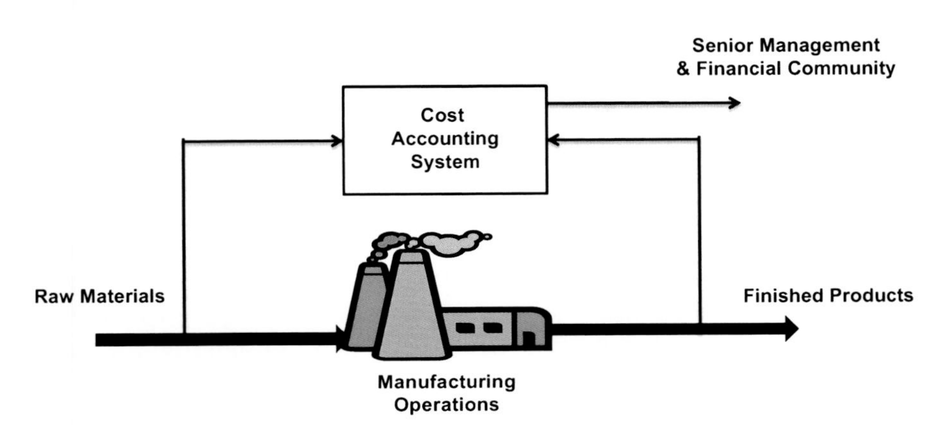

Fig. 2.8 Traditional cost accounting

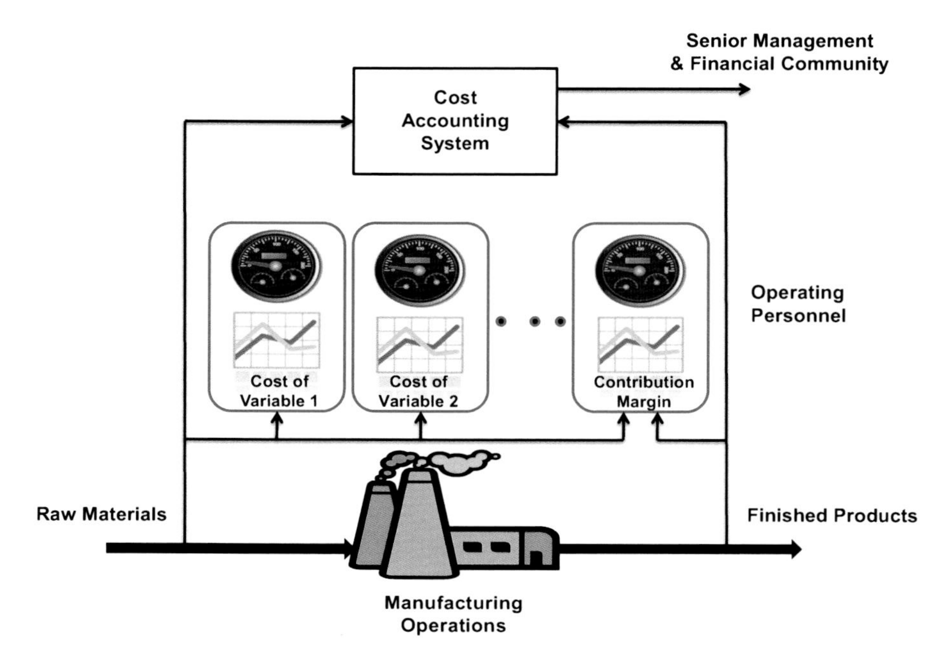

Fig. 2.9 Dynamic performance measures

2.5.1 Improving Chemical Production

Here is an example of a large chemical company, which was trying to improve its manufacturing performance. The company manufactured high value colorants and pigments, automotive and industrial coatings, and crop-protection agents. The area under consideration produced an intermediate chemical product, which was used in manufacturing many nylon products, such as carpeting and upholstery. The operation engineer for the production area was responsible for ensuring that the production rates were met, the plant ran reliably, and the products were produced according to specification, in terms of quality and cost.

The focus for improvements was on the distillation and oxidation areas, where the business strategy was to achieve the highest possible yield at the lowest possible production cost. The manufacturer understood that better control of raw material and energy usage could result in significant improvements in profit margin. In the past, the production management team relied on traditional "end-of-month" reports for reconciling unit and area profitability based on aggregated monthly raw material consumption and energy usage.

To determine cost-per-unit-product-made, production personnel would look at drum levels, raw material receipts, etc. daily but could only accurately reconcile once a month to compare how much raw material and energy was consumed versus how much product was produced. This approach added a significant amount of dead time and they could not realize and resolve production or raw material problems in a timely manner.

This inability to measure profitability in an appropriate timeframe extended to decision-makers at all levels of the organization. Operators, for example, were asked to watch the steam usage but they could not effectively monitor the resulting profitability from their consoles.

The Solution

The chemical company's culture of innovation and continuous improvement led the operation engineer to look into DPM as a possible way to improve its business performance. As a first step, they hired DPM consultants who conducted a top-down decomposition of the plant strategies, activities, and measures along with the functional hierarchy of the process areas. They developed models to calculate the performances, which were installed in the process control computer. DPMs were calculated in real time and for hourly, shift, and daily averages, which were displayed for the operating staff and their managers. The information was also stored as historical records.

There were two performance visibility tools. The first was a bar graph for process operators that resided on the control computer, which indicated real-time performance levels. The second performance tool was a number displayed on the computer screen of the managers every 24 h. This number was the ratio of raw material and energy used per pound of product made.

Using DPM, members of the team were able to measure in real time the results of the operational and engineering improvements they executed. If an area was burning too much fuel, for instance, it became very apparent, and the team was alerted immediately to the problem. Additionally, the system converted the use of raw materials to a dollar figure, which was displayed for the technicians. This made each technician aware of how his current actions were affecting the profitability, instead of waiting for an aggregate number at the end of each month.

The chemical company estimated that DPM have been a contributing factor in more than $2.2 million in annual profits, primarily attributable to cost savings. The profitability improvements in this production area convinced the company to expand the utilization of DPM to another nearby plant, where 100 % of the return on investments (ROI) was realized within 1 month.

2.5.2 Conclusions

The operating staff in a manufacturing plant keeps an eye on many process variables, such as temperature, pressure, flow/production rates, and product quality. However, looking at the cost of raw materials and energy is not so common. DPM allows them to observe those costs in addition to the usual information, such as process variables and the product quality. This makes the operating staff more aware of their contributions to the company's profitability and motivates them to fine-tune the process to gain additional benefits.

In many manufacturing plants, adding DPM does not require huge financial investment. Generally, no additional hardware is needed for systems that are controlled by computers. The process variables that are needed to calculate DPM are generally available since they are needed for the control of the manufacturing process. The additional displays can be built using existing hardware and software. However, some additional software may be needed for performing real-time calculations. The biggest expenditure may be in performing the feasibility study to ascertain its need and in training operating personnel to use it properly.

The benefits of DPM derive largely from a much faster feedback of cost and profitability, which focuses on overall process rather than individual control loops. Thus real-time performance measurement and reporting can drive a business toward optimal performance while adapting to changing market demands.

2.6 Process Characteristics

2.6.1 Instantaneous Response

A process with instantaneous response reacts immediately to a change in its load or its input. That is, its output follows a change in its input or load so quickly that for all practical purposes the process has no lag. When we change the position of the switch for an electric lamp the change is instantaneous. Similarly, when we open water tap in our home the flow changes instantly, which is roughly proportional to the amount of opening. Thus, in this case there is negligible lag between the action and its result.

2.6.2 Delayed Response

Most systems exhibit delays, which profoundly affect their behavior. Delays make systems not only sluggish but also make them less stable (oscillate) in certain situations. Delays are of two main types, the first is the dead time and the other is lag.

Dead Time

Dead time is a finite amount of time before an output responds due to a change in the input. When we open a hot water tap in our home, which has not been used for some time, there is a finite time before it delivers the hot water. This is the time it takes for hot water to flow from the tank to the tap. Until the hot water reaches the tap there is no perceptible change in the water temperature (Fig. 2.10).

When a newscaster on television interviews a correspondent located on another continent, we often notice dead time of a few seconds between the question asked and the response. This is due to the actual voice transmission time between the interviewer and the interviewee back and forth via satellite. When we are observing a star in the sky, which is many light years away, we are observing its status and position that was true that many years back. We have no way of knowing what its status or position is at the present time.

Suppose we want to drive a lunar rover on the moon, with a simple robot arm. An operator in an earth bound control station will control the rover and the robot arm using a joystick and observe its movements on a television screen. It takes more than a second for a signal sent from earth to reach

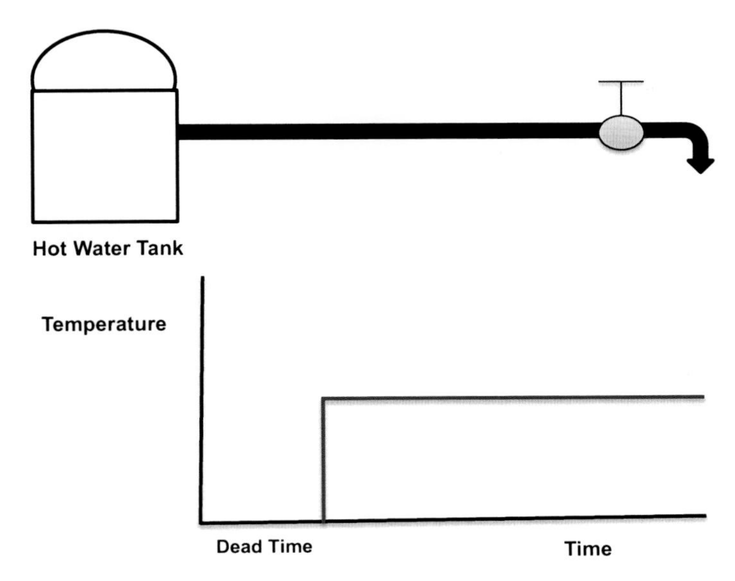

Dead time = Time taken by the hot water to travel from the tank to the faucet

Fig. 2.10 Example of a dead time

the rover on the moon and a similar amount of time for a signal to get back. Therefore, the operator will need to wait for over 2 s to see the results of a simple move made from the earth. If the operator acts hastily, she may make multiple moves before waiting for the results causing the arm to move beyond the target position and then retract it too much. This will cause a number of over and undershoots before the arm is in the right position.

Dead time is common in many industrial processes, where it is commonly referred to as transport delay. In many situations, those delays may be reduced if sensors are placed as near as possible to the process outputs.

2.6.3 Lag

First Order Lag

Lag is different from dead time, where a change in the input to a system causes the output to start changing almost immediately, but the full effect of the change in input takes some time to manifest. Lags are common in most systems. Increased pressure on the gas pedal, while driving a car, for example, does not cause the vehicle to gain the desired higher speed instantly, even though it starts to accelerate almost immediately. Attaining the desired speed takes time, which in this case depends largely on the power of the engine and the car's total weight (inertia).

In an open tank (Fig. 2.11), the water outlet flow is dependent on the tank level and the amount of opening of the outlet valve. If we keep the valve in a fixed position, then the outlet flow depends only on the height of the water in the tank. The higher the level in the tank the more is the outlet flow. Let us assume that the system is in equilibrium, that is the flow in and out are equal and the water level in the tank is steady. Now, if the inlet flow increases suddenly, the level will start to rise resulting in an increase in outlet flow. Eventually, the inlet and outlet flow will be equal and then system will be in

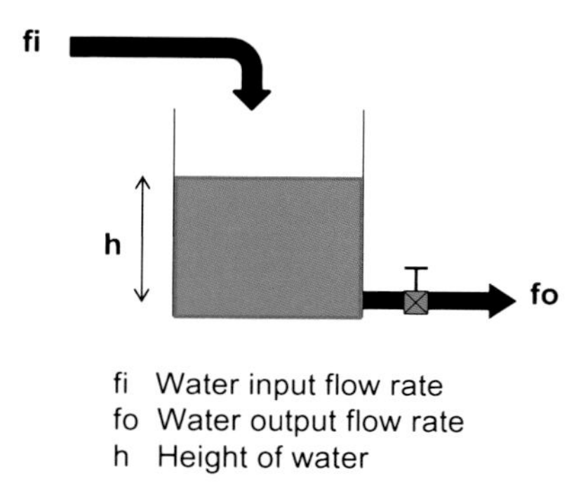

Fig. 2.11 Example of a first order lag

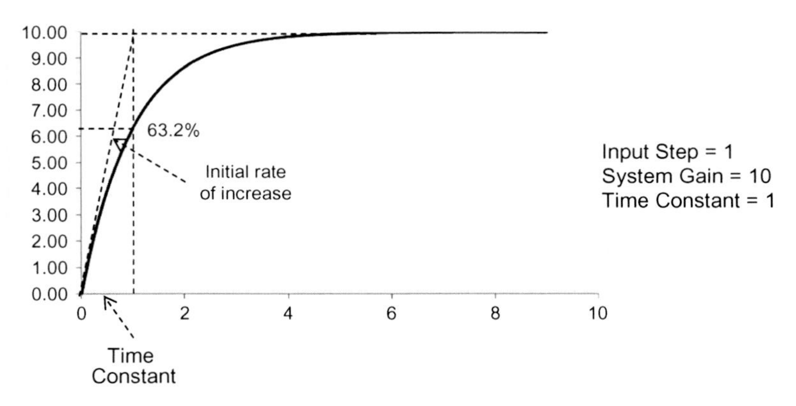

Fig. 2.12 Response to a first order lag

equilibrium again, but at a new and higher level of water in the tank. This is a typical example of first order lag (Fig. 2.12) where a step change in the input leads to gradual change in output.

The time taken to reach nearly two-thirds (63.2%) of the final value is defined as the time constant of the system, which in this example is 1 min. Time constant may also be defined as the time taken for the output to reach the final value had the original rate of increase been maintained. One thing to remember here is that the time constant of a system does not vary with the magnitude of the disturbance. Time constants are found in many different kinds of systems, such as resistor–capacitor circuits in electrical and electronic systems, speeding up and slowing down of motors, and heating and cooling down of lakes with the change in atmospheric temperature. Lags are common in industrial systems and are even more common in social, political, and economic systems.

Strictly speaking, time constant applies only to simple systems with first order lag as in this example of a single water tank. For a more complex system with multiple lags, the system may be resolved into a number of simple systems in series. There, the time constant of the whole system is a combination of the time constants of its individual components. However, time constant is a useful measure because it indicates the speed of a system's response.

Second and Higher Order Lags

A system with second order lag may be considered as two systems with first order lag in series. That is, the output of one system feeds to the next like two water tanks in series (Fig. 2.13). A step increase in input to the first tank

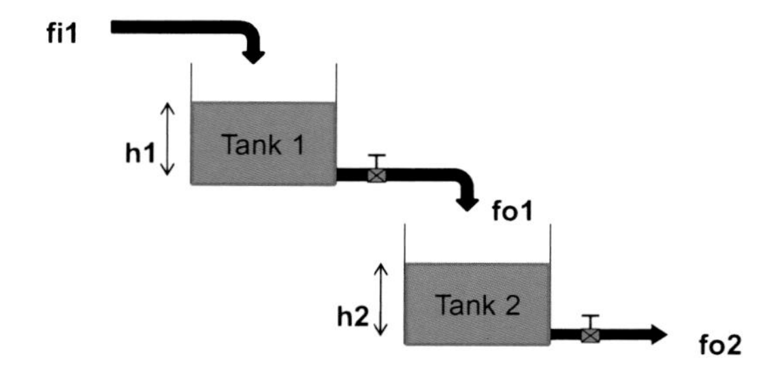

Fig. 2.13 Example of a second order lag

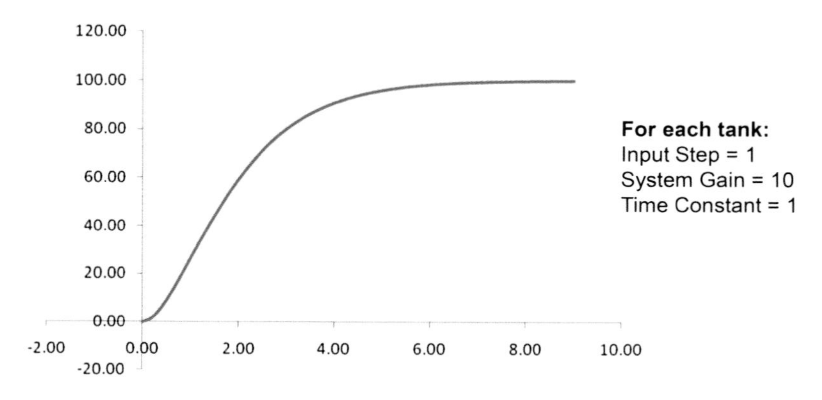

Fig. 2.14 Response to a second order lag

will cause an increase in its outlet flow but the output of the second tank will change more slowly because it did not get a step input, but a gradual rise in input flow. The output flow from the second tank will then be an “S”-shaped curve (Fig. 2.14).

2.6.4 Simple and Cascaded Control Loops

A water tank, which supplies hot water, is heated by steam that flows through a coil in the tank (Fig. 2.15). The flow of steam through the coil needs to be controlled in order to maintain a constant temperature at the hot water outlet. In a simple arrangement, the flow of steam is controlled by a single controller that monitors the hot water outlet temperature, where the desired temperature is set manually (Fig. 2.15a).

The controllability can be improved by a hierarchical arrangement (Fig. 2.15b), where instead of directly controlling the steam flow based on the hot water outlet temperature, the flow of steam is controlled by a flow

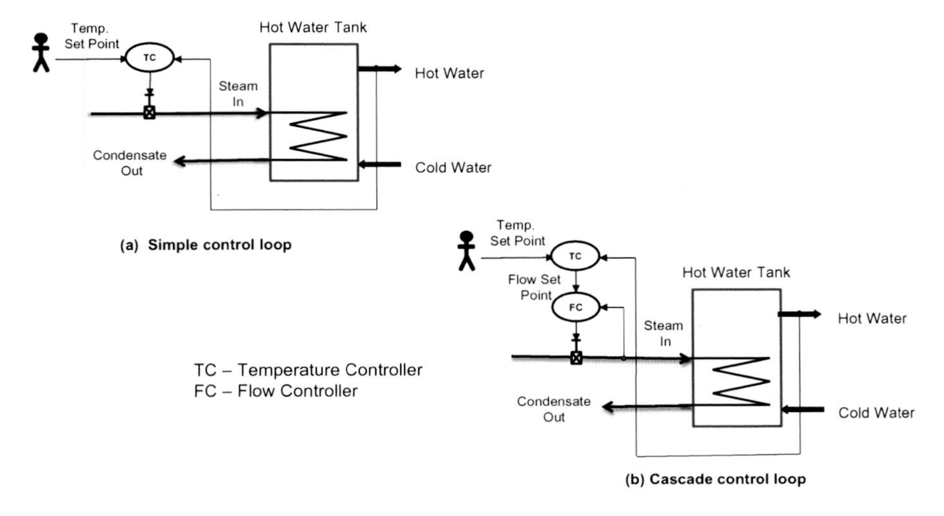

Fig. 2.15 Hot water temperature control

controller, whose set point is set by the temperature controller. The temperature controller's set point is set by a manual operator as before. Cascaded control is the common term used in the industry for this type of hierarchical loop structure, where the output of the controller in the outer loop is the set point of the controller in the inner loop. The outer loop is often called the primary or master, while the inner loop is called secondary or slave. Cascaded loops are often used in the industries as they work faster in correcting disturbance, such as the change in the steam pressure at the inlet.

2.7 System Interactions and Hierarchies

Systems interact with each other in peer-to-peer fashion and in hierarchies. Automobiles are supposed to interact with each other in a peer-to-peer fashion by following simple traffic rules, such as driving on the right side of the road and at an intersection with no traffic lights allow vehicles on the main road to pass before those on a side street.

Rigid hierarchies are the norm in most engineered systems because they are expected to behave in precise and predictable ways. These hierarchies may be viewed as extensions to the concept of cascaded loop discussed in the previous section. For example, an aircraft has a number of functional modules, such as structural, propulsion, navigation, and control that have to work in a unified fashion under the directions of pilots and flight engineers (Fig. 2.16). Each of

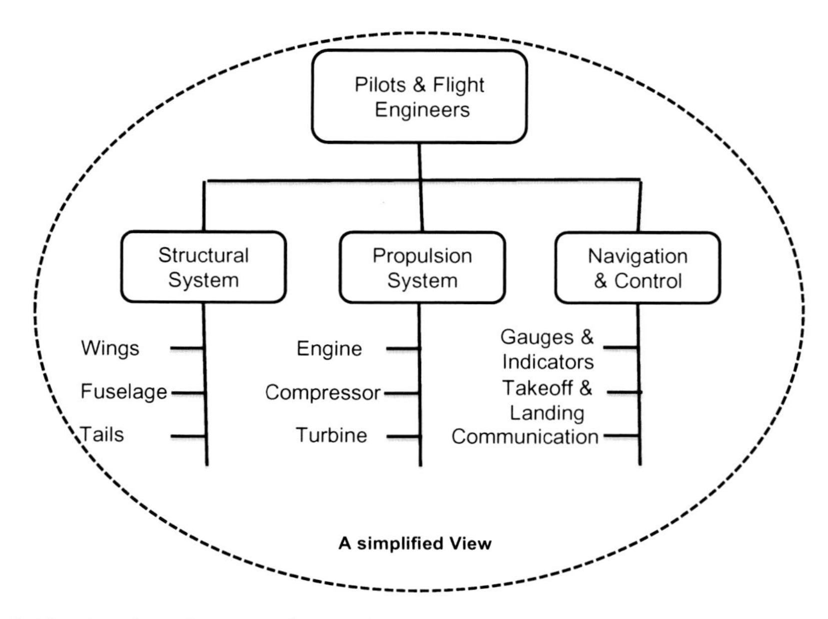

Fig. 2.16 An aircraft system hierarchy

these systems has a number of subsystems, such as engine, compressor, and turbine for the propulsion system, which again have to work in a unified fashion for the aircraft to fly. Thus, there is a distinct hierarchy where commands flow from higher to lower levels while status information is passed on from lower to the higher levels. Similar is the situation in many other engineered systems, such as pharmaceutical manufacturing or food processing plants, where strict quality requirements of products necessitate exacting procedures with rigid hierarchies.

Centralized supervision and control is the hallmark of a traditionally designed system. However, in order to augment robustness and reduce vulnerabilities of large and complex systems designers are increasingly looking for decentralization of decision-making and control. That is discussed in detail in a later chapter.

2.8 Simple Feedback Control: Strengths and Limitations

Feedback control, as described in this chapter, is used extensively in engineered systems. They are simple and cost effective. In many cases, they remove drudgery and save labor. In a simple feedback control loop, the output of a controller at any instant depends on the error, which is the difference between the set point and the feedback. It is true that integral action of a controller is based on the sum of immediate past errors, but does not entail keeping a model of the process being controlled. Feedback control is robust for simple applications where the action of a controller is apparent almost immediately. In the example of driving an automobile, the action on the brake is immediate in slowing down the vehicle. In controlling the delivery of oil to a burner, a small change in the opening of a valve has an immediate effect on its flow.

However, simple feedback control does not work well with processes that have slow response, time delays, experience frequent disturbances, or have multiple process inputs and outputs that interact with each other. More advanced control functions are needed to control those processes effectively, which are discussed in later chapters.

Loop diagrams are good at showing interrelationships within a system or between systems but are not good enough in showing their dynamic behaviors. Thus, they cannot predict the current or future behaviors of a system. For that, we need to observe the behavior of an actual system or simulate it by using a modeling software package (see chapter on modeling and simulation).

2.9 Engineered Systems Since the Days of James Watt

Engineered systems have come a long way since the days of James Watt. They are all around us, be they automobiles, airplanes, radio, televisions, televisions, computers, internet, electric supply, or water supply systems. We are dependent on their proper working to serve us in our daily lives.

With the development of engineered systems came greater understanding of how they should be designed and used. James Clerk Maxwell was the first person to analyze mathematically the workings of the Watt Governor. He was followed by many scientists, mathematicians, and engineers who laid the foundations for systems science based on their understandings of engineered systems. Lately, however, persons in other disciplines such as psychology, medicine, social science, and economics have found the benefits of applying system knowledge in their respective fields. In the next chapter, we will discuss the application of system knowledge in these disciplines.

Why Do I Need to Understand the Dynamics of Engineered Systems?

Scientists and engineers who design, build, and maintain these systems need deep understanding in order to design them properly and to optimize their functional behaviors. For the rest of us who use these systems there may not be the need for such detailed knowledge, but an appreciation of their behavior patterns and their reasons can help us in using them optimally. Additionally, as stated before, the dynamic behavior of engineered systems provides analogies for the workings of others, such as political, economic, social, and biological systems. This helps us better understand and appreciate the workings of natural, social, and other such systems and offers a path for their detailed study.

Is Control a Part of a System?

This question often comes to mind when looking at system behaviors. In engineered systems, the controller and the controlled (process) are generally distinct entities. Thus, a cruise controller in an automobile or an autopilot in an aircraft is separate from their engines and other devices that make them run. Therefore, they are separate subsystems, whose behaviors may be studied in isolation, especially when the controllers are yet to be developed or are not active. However, when studying the behavior of an automobile in cruise control or an aircraft on autopilot, the whole system that includes the controller and the controlled need to be viewed as a single system.

Right Information at the Right Time

Earlier in this chapter, there is a case history of a chemical company, where the continuous measurement and display of the profitability of one of its plants helped the operators to better optimize its operation and thus reach greater levels of productivity. This concept may be applied in many other situations for similar benefits.

Donella Meadows, in her book "Thinking in Systems," mentions the varying consumption of electricity by two identical sets of households in Amsterdam in The Netherlands, where one set of homes had their electric meters down in their basements, while the other had theirs in their front halls. These meters had horizontal metal wheels that were clearly visible, whose turning speed depended on the rate of consumption of the electric power. Those houses with meters in their front hall were found to use lesser amounts of power than those that were tucked in their basements. This indicates that instant visibility has the effect of reducing electric consumption. Today, many fuel-efficient vehicles, such as the Toyota Prius display fuel consumption rate prominently on their dashboards thus encouraging economic driving habits.

A Closed-Loop System May Work Well Even When There Are Inaccuracies in Measurements or Its Dynamics Are Poorly Understood

In the example of a home heating system, if the temperature sensor is inaccurate, the set point may have to be altered. For example, if the sensor always reads 68 °F when the actual temperature is 70 °F, then we need to adjust the set point two degrees lower than the required temperature to maintain the desired temperature. Here the word "always" is important, that is, the inaccuracy is constant and does not vary with time or in some random fashion. The same concept may be used in an economic situation like unemployment. If the unemployment data for a country is inaccurate and is always underreported, the effect of the actions taken to reduce unemployment may still be measured with reasonable accuracy so long as the inaccuracy in the measurements remains constant. However, in other applications, such as an error in measuring the altitude of an aircraft the effect may catastrophic as it lands unless the pilot knows the consistent error of the altimeter.

The advantage of feedback control is that it works well most of the time even when the dynamic working of a system is poorly understood. A system may have multiple variables that affect its outcome and whose interactions are not well defined, such as for a steel rolling mill where the temperature of a heated slab, clearance between each pair of rollers, and the speed difference between pairs of rollers affect the thickness of the rolled steel sheets (Chap. 1). However, in a closed-loop situation the thickness can be closely controlled when only one of these variables is manipulated in real time based on the feedback from the thickness gauge for the finished product.

Thus, good feedback leads to self-correction. With proper feedback, a poorly understood system can be better managed than a well-understood system with little or no feedback.

Effects of Lag and Dead Time

Dead time and lag are inherent in most systems. Often they are considered a nuisance that unnecessarily slows the systems. However, by closer examination of such delays in engineered systems one gets greater understanding of their true effects, which range from poor controllability to oscillations. This makes us understand that better controllability in a system may be achieved by merely reducing the delays in that system, which will be discussed later in a greater detail.

3

Political, Social, and Biological Systems

Abstract The studying of behavior patterns of political, social, and biological systems began in earnest in the middle of the twentieth century. The underlying principles of closed-loop control in mechanical and electrical systems had been studied earlier by many professionals in other disciplines. Professor Jay W. Forrester of MIT led the study of the behaviors of social and business systems based on the principles of feedback and closed loop. What he found was a large measure of commonality in the behavior patterns of the systems in these various disciplines.

This chapter is an introduction to the behaviors of political, economic, social, and natural systems. It starts with the history of the study of such systems and shows how they are depicted by causal loop diagrams. There is discussion of biofeedback, dead time and lag, and system hierarchies. That is followed by a case study on urban youth violence. The chapter concludes with a discussion on the advantages and limitations of causal loop diagrams and with an example of system dynamic approach in our daily lives.

3.1 Jay W. Forrester and the History of System Dynamics

The person who spearheaded the study of the dynamics of business, economic, and social systems was Jay W. Forrester, a professor in the Sloan School of Business at the Massachusetts Institute of Technology (MIT). Forrester came to MIT in 1939 as a graduate student in electrical engineering. He first worked as a research assistant under Professor Gordon Brown, the founder

A. Ghosh, *Dynamic Systems for Everyone*,
DOI 10.1007/978-3-319-43943-3_3

of Servomechanism Laboratory in MIT and did research on feedback control mechanisms in engineered systems.

At the end of World War II, he pioneered the use of digital computers in an aircraft flight simulator for the US Navy. In 1947, he became the director of MIT's newly founded Digital Computer Laboratory. There he was instrumental in the creation of WHIRLWIND I, MIT's first general-purpose digital computer. During the design and construction of the computer, Forrester was one of the inventors of random-access magnetic core memory, which became the industry standard for the next 20 years until it was replaced by solid-state memory chips.

3.1.1 System Dynamics

While managing the computer laboratory along with his work on simulation and feedback control of engineered systems, Forrester gained an appreciation of the difficulties that executives face in managing businesses and other organizations. He concluded that social systems are much harder to simulate and control than physical or engineered systems. Thus, the management problems in an industry are more formidable than engineering or technical challenges.

In 1956, Forrester switched his focus from engineering to management by accepting a professorship in the newly formed MIT School of Management, which was later renamed as the Alfred P. Sloan School of Management. There his aim was to use his knowledge of the behavior of engineered systems to solve management problems in various corporations. He got involved with General Electric Company (GE) because of a problem with their employment cycle, where there were alternate surplus and scarcity of workers. The management at the GE plant looked at incoming orders and inventory build-up to determine future staffing needs. When they found a discrepancy, they reacted accordingly. The problem with this system was that it had built-in delays, which Forrester was able to show by simple simulation of GE's decision-making structure. It could take many months to satisfy hiring needs and many more months to train the new hires. Thus, the employment and business cycles of GE were uncoordinated.

During the late 1950s and early 1960s, Forrester's team, including his graduate students, created the first system dynamics-oriented computer-modeling package called SIMPLE. That was followed by DYNAMO an improved version of the earlier package. In 1961, Forrester published the first, and still classic, book in this field titled *Industrial Dynamics*.

3.1.2 Urban Dynamics

In 1968, Forrester met John Collins, the former mayor of Boston and the two began to engage in regular conversations about the problems that cities face and how system dynamics may be used to address them. The collaboration resulted in a book titled *Urban Dynamics*, along with a model with the same name, which addressed the common problems facing a city. The model allowed the simulation of alternate policies to improve city management. It showed that the traditional approach of a city government, where different departments, such as housing, sanitation, and welfare work in isolation, are relatively ineffective in solving the problems of modern cities. The model was and still is quite controversial as it demonstrates that many well-known urban policies are either ineffective or have negative impacts, while effective policies are often counterintuitive. The strength of Forrester's work lies not in any specific model but the insight he offers in solving urban problems. His works inspired the development of a well-known video game called SimCity.

3.1.3 World Dynamics

In 1970, the Club of Rome, an organization that is concerned with the future of humanity and the planet, asked Professor Forrester whether system dynamics could be used to address the predicaments related to the fast growth of population in this world, such as the availability of resources and the disposal of pollutants. That question resulted in the research and publication of a book titled *World Dynamics*, which showed a possible collapse of the world's socio-economic system sometime in the twenty-first century unless the demands on earth's resources are reduced. *World Dynamics* was later revised by a group of his former students led by Dennis Meadows.

Professor Forrester has also been involved in the creation of a national economic model, which, though not fully finished, shows that economic cycles of 40–60 years are a feature of a capitalistic economy.

3.1.4 System Dynamics as a Part of Children's Education

For a number of years, Professor Forrester has been focusing his attention on extending the teaching of system dynamics to children from kindergarten through high school, with the expectation that future decision makers would be able to take a systemic view of the problems they face, thus greatly benefitting the society.

Well-Known Quotations from Jay Forrester

- Biological and social problems are manifestations of the underlying nature of complex systems.
- When someone tries to change one part of a system, it pushes back in uncanny ways, first subtly and then more forcefully, to maintain its own implicit goals, for example:
  - A dieter's body will seek to maintain its present weight by producing cravings for fattening food.
  - In a corporate reorganization, employees tend to carry on their old ways thus circumventing the new system.
- Modeling growth and resistance in biological or social systems requires nonlinear calculus, which even the most gifted mathematicians are incapable of solving in their head, yet corporate and political leaders continue to make decisions based on their mental models.
- Many of the problems the world faces today are the results of short-term measures taken in the last century.
- Most of the pressing problems facing humanity today will elude solution until a new generation, familiar with computer models, gets into leadership role.
- The problems of most businesses are not brought on by the competitors or market trends, but are the direct results of their own policies.
- The great challenge for the next several decades will be to advance understanding of social systems in the same way that the past century has advanced understanding of the physical world

3.2 Causal Loop Diagram

Non-engineered systems, which include political, economic, social, and natural systems, are commonly depicted by causal loop diagrams. These are somewhat similar to block diagrams. They both show actions and their feedback reactions but their points of emphasis are different. An engineered system generally works within a narrow band, that is, it performs at or near its desired goal or set point. For example, a home heating system maintains the room temperature around its desired value while facing disturbances like opening and closing of doors and changes in the outside temperature. A Watt Governor

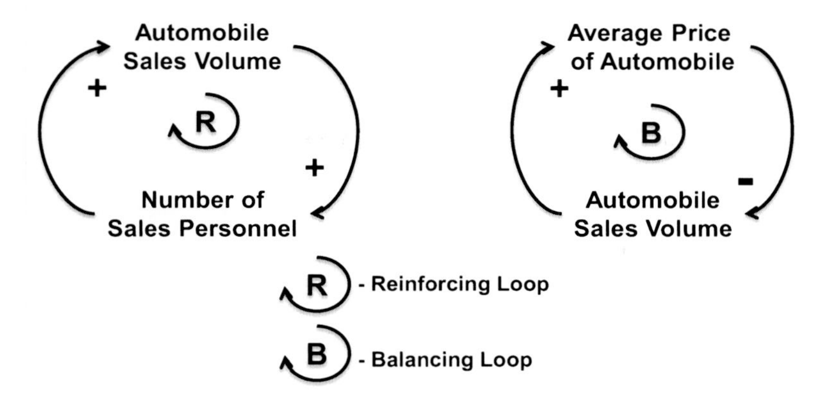

Fig. 3.1 Causal loop diagrams

is supposed to keep a steam engine running at the set speed in spite of the change in load or fluctuations in its boiler pressure. Thus in engineered systems, the set point and the control functions loom large. However, in social, political, economic, biological, and other such non-engineered systems, the different variables influence each other and in most cases have certain desired objectives rather than fixed set points.

In causal loop diagrams, control functions are not depicted as distinct entities, as they are in control loop diagrams. Instead, the focus is on the depiction of system variables and their influences on each other. In addition, in causal loop diagrams, there are no geometric structures to specify functions or subsystems. Variables are specified textually or pictorially while their interrelations are shown by arrows, as illustrated in the following examples (Fig. 3.1).

The figure illustrates two scenarios for an automobile dealership. The diagram at the left shows a positive correlation between sales volume and the number of sales personnel. As the sales increase, the number of sales people is increased to serve the additional customers. Again as the number of sales personnel is increased the sales volume increases even further. Therefore, these two variables reinforce each other, as is shown by the pair of arrows. Each arrow shows the direction of influence along with a positive (+) or a negative (–) sign. A positive sign indicates that when the variable at the tail end of the arrow changes, it influences the variable at the head to change in the same direction. Conversely, a negative sign indicates that when the variable at the tail end changes it influences the variable at the head to change in the opposite direction. Conceptually they are similar to positive and negative feedbacks in engineered systems. The positive and the negative signs do not indicate the increase or decrease of the pair of variables, as they may go in either direction, nor do they indicate the desirability or undesirability of their directions of change.

The loop at the right in the same figure shows the interrelationship between the average prices of automobile with its sales volume. As the average price is raised, perhaps by giving fewer discounts and other price incentives, the sales volume goes down. There the inverse influence is indicated by the negative sign. However, as the sales volume goes down it influences the management to reduce the price to increase the sales volume. Therefore, the automobile sales volume influences the average price in the same direction, which is shown by a positive sign. It is important to note that in these examples there are only two variables in each loop. There is, however, no limit to the number of variables in a loop diagram so long as they are related to each other.

The loop at the left, where the variables reinforce each other will lead to increased growth or collapse, is called a *reinforcing loop*, which is depicted by an "R" in the center, with an arrow showing the direction of the influence. In the loop at the right the two influences counteract (negative feedback) thus creating a balance. Therefore, this is called a *balancing loop*, which is depicted by a letter "B," with an arrow showing the direction of the influence. If in a loop there is no negative sign, then it is a reinforcing loop. Additionally, a reinforcing loop may have an even number of negatives, because as we all know, multiplying two negatives makes a positive. If the total of negative signs in a loop is an odd number, then that is a balancing loop. In general, a reinforcing loop creates growth along with the possibility of instability, while a balancing loop provides the needed stability.

In this example, the two loops are related, so they may be combined into a single diagram (Fig. 3.2). Time delay, which is common in many systems is depicted by a set of parallel lines (=) as shown in this figure.

Even though there has to be a minimum of two variables to form a loop, often more than two variables interact. This is illustrated by the example of the effect

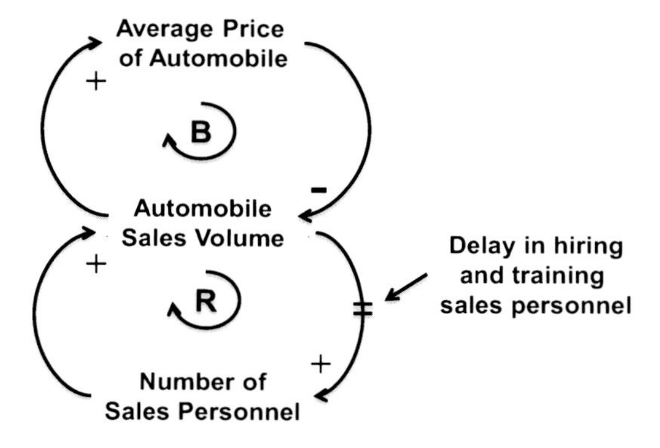

Fig. 3.2 A causal loop diagram with multiple loops

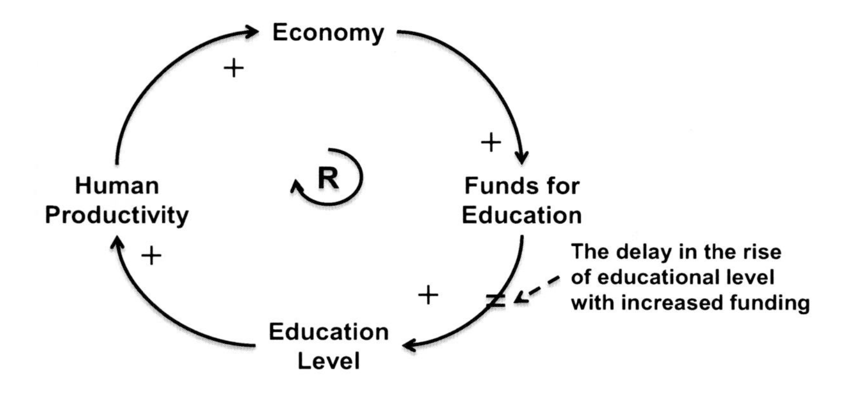

Fig. 3.3 A causal loop with multiple variables

of education on an economy (Fig. 3.3). Four variables are considered here: the economy, funds for education, level of education of the citizens, and human productivity. These four variables exert positive influence on each other: a rising economy increases the funding for education leading to a higher level of education for its citizens. This leads to higher productivity level of its citizens thus driving the economy to an even higher level. At the same time, the diagram also illustrates that a falling economy will cause reduction in educational funding, leading to a decline in educated work force, which will drag down the economy even further.

It should be noted that a variable specified in a causal loop diagram should preferably be a noun or noun clause and not a verb. For example, "Economy" and "Education Level" are variables that are stated correctly, while "Economy Going up" and "Education Level Going Down" are not. A set of guidelines for drawing control loop diagrams is provided in the appendix.

Causal loop diagram is an important tool for describing system interactions. By drawing these diagrams individually or as a team, we get a rich array of perspectives on what is happening in a system. Thus, we are able to find ways to make the system work better. By understanding the connection between the average price of motor vehicles and the number of sales personnel needed, for example, management can make an informed decision to increase or decrease sales incentives and to plan on the number of sales personnel to be employed.

Causal loop diagrams have been found to be useful in studying systems in various disciplines, because by using simple basic elements, one can describe from simple-to-very complex interactions. They are particularly useful for finding and conveying the dynamic behavior of systems, which often elude simpler mental models. Often, software packages for modeling and simulation support the creation and display of causal loop diagrams.

3.3 Feedback

Feedback is integral to a natural ecological system. As described in Chap. 1, the number of predatory animals in a given geographical area is kept under control by the number of animals available for prey. Similarly, the number of animals for prey is controlled by the number of predators, leading to an oscillatory but balanced system.

The price of a consumer product is generally determined by its demand and its supply, which act as feedback to each other. Increase in demand for a product leads to its scarcity, which in turn causes its price to rise. Manufacturers perceiving the opportunity to increase profits, boost production leading to price stabilization. Of course, there is always the possibility of overproduction leading to a significant fall in price, followed by cuts in production causing scarcity. This, rise and fall in price is termed *oscillation*, which is quite common in many systems and will be discussed in detail in the next chapter.

Feedback is very important in the social context. A smiling baby is feedback to the mother that everything is fine, while crying signals that something is wrong, be it hunger or some other discomfort. Friends or married couples provide feedback to each other indicating affection, indifference, or outright hostility. Feedback is effective in a closed-loop situation, such as a mother trying to understand and resolve the problem when her baby cries or a couple trying to understand and resolve their mutual problems.

In a functioning democracy the periodic elections, the free press, and other media provide much-needed feedback for the administrators. In the past, benevolent kings used informers to gauge public opinion to shape their policies. So did some dictators with malevolent intentions. Without feedback, they would have had little idea about the effectiveness of their policies and thus would have been at a loss in shaping their present and future course of actions. We use the term "losing touch with public," when a politician gets too busy with administration and finds little time to interact with the public and gauge its opinion. That also raises the point about the quality of feedback and the possibility of its distortion, which often happens when the person in authority is shielded from the public and is presented with distorted views by the surrounding courtiers and underlings.

3.3.1 Biofeedback

Here is an example of a person who takes the functioning of his mind, body, and spirit very seriously. He is a young professional, named Justin, who while waking up in the morning takes off the headband that he has been wearing all night that measured his brainwaves and stored them in his personal computer. These recordings produce bar graphs of his periods of deep sleep, light sleep, and rapid eye movement (REM). Still half asleep, he goes to the bathroom and steps on a digital scale, which communicates with his personal computer to record his weight. At the breakfast table, he takes a photo of the food on his plate with his smart phone, which calculates the calories in the meal and stores it in a database in his computer. He puts on an armband blood pressure monitor and a sensor for monitoring the heart rate and then bikes off to his office. His bike is equipped with probes for tracking the time, mileage, and the force applied on the pedals. After getting into his office, he retrieves these data from his bike to his mobile phone and then transmits them to his personal computer. He laments that no sensor is yet available to analyze his urine readily.

Justin's weight, exercise habits, caloric intake, sleep patterns, and other vital signs like heart rate and blood pressure are quantified in spreadsheets and graphed in his personal computer and, of course, he can retrieve all this information on his mobile smart-phone as needed. Based on the data, he makes his daily decisions, such as the quantity and type of foods to eat, the number of cups of coffee to drink, the intensity and the duration of physical exercise, and the number of hours he should sleep. Thus, he tries to optimize his mental and physical performance on a daily basis. Justin believes in treating his body as a whole, that is, as a single interconnected system, where different organs work together in a harmonious fashion, for most of the time, to keep it healthy and active.

Justin's case may be an extreme example of the use of personal biosensors and their feedbacks but an increasing number of people are using them, as they become more user friendly and cost effective.

3.3.2 Dopamine Treatment for Parkinson's Disease

In a normal person, the unlimited pursuit of pleasure is repressed by its possible adverse effects. Thus, there are two loops, one reinforcing pleasure-seeking activities, the other balancing that with the caution or fear of their negative consequences (Fig. 3.4). However, recent researches in Parkinson's disease show that patients without medication exhibit an inability to learn from positive or

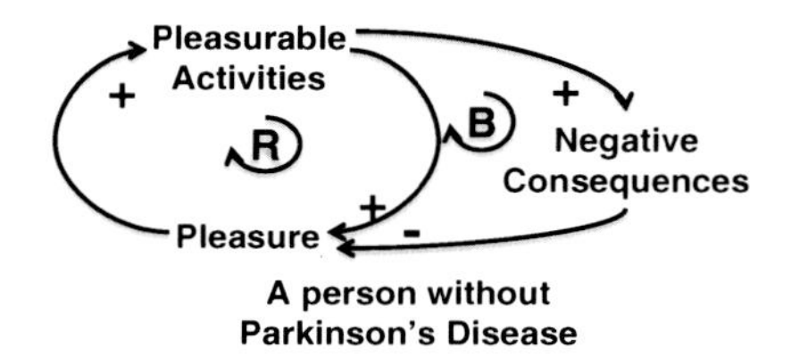

For a person with Parkinson's disease:

With no dopamine treatment -- the connection between 'Pleasurable Activities' and 'Pleasure" gets weak, leading to depression

With dopamine treatment -- the connection between 'Negative Consequences" and 'Pleasure" gets weak, leading to impulse control disorder

Fig. 3.4 Effect of dopamine on a person with Parkinson's disease

rewarding outcomes, while their sensitivity to negative or punishing outcomes stays normal. Thus, their reinforcing loop for pleasurable activities is impaired leading to depression. That is not surprising, as scientists have long known that patients lose most of their dopamine producing function, where, dopamine, a neurotransmitter, helps controlling the brain's reward and pleasure centers.

This is in direct contrast to what happens to most patients when they start dopamine treatment, a standard therapy for treating the physical symptoms of the disease. A patient's ability to learn from positive outcomes improves, while the ability to learn from negative outcomes, which had previously been normal, is reduced. Thus, the pleasure-reinforcing loop gets back to near normal, while the balancing loop that suggests caution gets weaker. That leads some patients to develop impulse control disorders, which may manifest as pathological gambling, hyper-sexuality, alcoholism, compulsive eating, and shopping. Thus, the ability to test the effects of medication in early onset of Parkinson's disease may provide additional insight into the impact of dopamine loss on cognition and behavior. It may also pave the way for identifying which Parkinson's patients are most likely to experience dopamine-related feedback problems so they may be treated with alternate medications.

3.3.3 Know Thyself

In the Indian philosophy there is a saying—"Atmanam Vidhi." In Greek, it is "Gnōthi Seauton." Both of these translate as "Know Thyself." This is sage advice when deliberating on biofeedback. A human body is a very complex system, more complex than any man-made system built so far. Knowing this system, even partially, is the basic objective of any biofeedback process. Most of our bodily functions both physical and mental work automatically, without our knowledge or active intervention. However, we can gain some level of control over those functions, when we are able to understand and measure them.

Medical practitioners everywhere rely on biosensors, such as those that measure heart rate, blood sugar levels, and brain activities. We use biosensors when we take our body temperature or when we look on a scale to check our weight. Biofeedback devices go a step further by monitoring and recording our bodily conditions and reporting them back in real time and storing them for future use. They commonly include sensors for blood pressure, heart rate, skin temperature, sweat level, muscle tension, and brain activity. Modern gym equipment like treadmills has biosensors that tell heart rate and calories burnt, which motivates their users to use them more often.

An interesting result of being able to read our body's systems is that we can control more muscles and other organs than were previously thought to be controllable. We can see what happens when we are able to gain control over such things as our own heart problems and muscle tensions.

Biofeedback is often aimed at changing habitual reactions to stress that can cause pain or disease. Many believe that we have forgotten how to relax. Feedback of physical responses such as skin temperature, sweat level, and muscle tension provides information to help us recognize a relaxed state. The feedback signals also act as a kind of reward for reducing tension. It is like a teacher whose frown turns to a smile when a student finally gives the correct answer.

The value of a feedback signal as information and reward is even greater in the treatment of patients with muscle paralysis. With these patients, biofeedback is a form of skill training like learning to play a game. When the patient tries to stimulate an affected muscle, he or she can watch the feedback from the machine, thus enabling the patient to try various different ways to activate it. Perhaps just as important, the feedback convinces the patient that the limbs are still alive, which often encourages the patient to continue the efforts.

The benign effects of biofeedback highlight the importance of feedback in real time. To be able to measure some bodily functions sometimes in a hospital, in a doctor's office, or in one's home is not enough. The ability to measure and observe the bodily functions continuously gives it an edge over traditional diagnostic procedures. Thus, biofeedback has the potential to reduce medical expenses by fostering healthier lifestyle and by facilitating early diagnosis of latent and impending ailments. New developments in biofeedback sensors that are less invasive and more cost-effective will definitely increase their usefulness and wider acceptance.

3.4 Dead Time and Lag

In the last chapter, we discussed the effects of dead time and lag in engineered systems. We define dead time as the amount of time between an action and the start of its manifestation. For example, if I mail a package to a friend, there is a definite delay of a few days or more between my posting and her receiving it, which is a dead time. Lag, on the other hand, is the time between an action and a significant manifestation (63.3 %) of its full effect. If a manufacturer, for example, reduces the price of a popular product, the resultant increase in sales may become apparent immediately but the full effect of the price reduction may lag by a few months or more, because it will take time for that information to get around.

Dead time and lags are quite common in our everyday life; all social, political, economic, and natural systems exhibit them. In fact, it would be quite unusual to find a system that does not have any dead time or lag. When I send an instruction to my bank to transfer money to a creditor's account, the time it takes depends on the dead time in the system. In the old days, it would have taken days for the information to get from one bank to another. Nowadays, with computers and instant messaging it may take only seconds although if human intervention is required for the transfer, then that may still take a day or more, depending on factors, such as the time of the day and the day of the week of the request.

In real life, we encounter many lags in series. For example, getting a permit to build a manufacturing facility in a certain location may require environmental clearance, permits from the zoning board, and the like. Often a series of lags add up to make the system more sluggish.

Large systems have inherent delays associated with them. The more bureaucratic an organization is, the more is the time delay for getting anything done. However, bureaucracy is not the only thing to blame; delays are inherent in any system. For example, it will take years to produce a better-educated workforce if we improve our school systems today.

Dead time and lag are often considered as pure nuisance, which lead to greater inefficiencies in the working of a system. However, their effects go much deeper than what is apparent. They have significant effects on system behaviors, as we will see later.

3.5 System Hierarchy

Hierarchical and peer-to-peer relationships are common between a system and its subsystems. In a hierarchical relationship, some aspects of a system may be controlled by another system at a higher level or may be dependent on the actions of systems at a lower level. In the last chapter, we discussed system hierarchy in engineered systems. That concept can be extended for the others, such as political, economic, and social systems.

In a large organization, such as a bank or a corporation, there are well-defined hierarchies, where each branch manager has duties and responsibilities for serving their clientele while the central office has the responsibilities for specifying common rules and rates. The branch manager in turn depends on her/his staff for the proper execution of the day-to-day activities of the branch. Within a biological system like a human body, each individual organ like heart, lungs, and individual muscle perform its particular function, but their actions are coordinated by the brain, which acts at a higher-level.

If we are focusing on a forest ecosystem, which consists of a range of plants, animals, and microorganisms along with the nonliving physical factors of the environment, it would be futile to look at individual trees or animals (Fig. 3.5). Similarly, if we are studying an urban traffic system, it would not help us if we focused on individual motor vehicles or their subsystems like engine, transmission, and brakes. Doing so would add unnecessary detail and complexity and would lessen our understanding of the problem we are primarily interested.

While focusing on a level in a hierarchy, it is prudent to take into account the main interactions in that level and the levels immediately above and below.

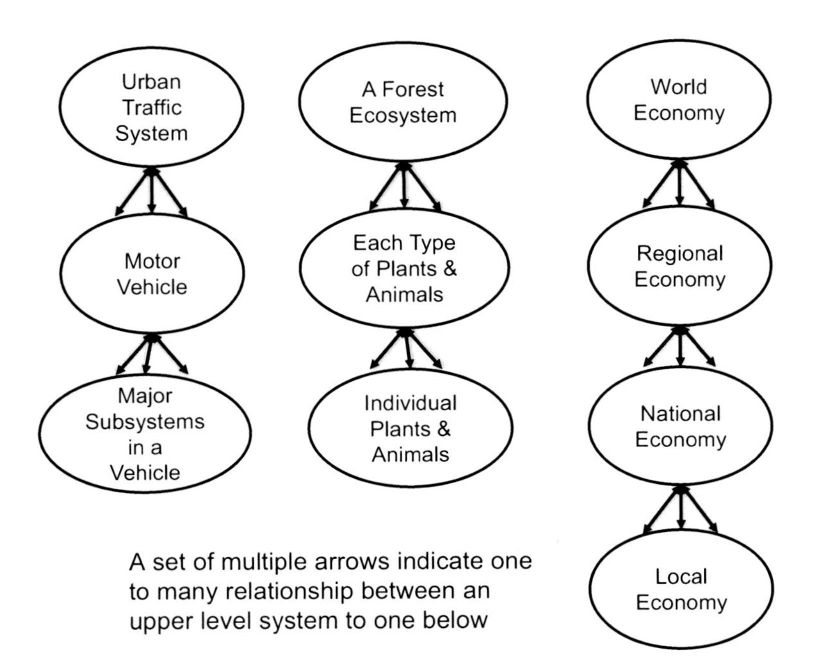

Fig. 3.5 Examples of system hierarchy

3.6 The System Dynamics of Youth Violence: A Case Study

Youth violence has become endemic in certain neighborhood of Boston as it has in many major cities in the USA and around the world. Pondering on the root cause of this problem, community leaders initiated the Youth Violence Systems Project (YVSP), led by Emmanuel Gospel Center in Boston in partnership with a number of community-oriented organizations. The project included people from a wide range of organizations and backgrounds, such as individual community members, community groups, church members, and public health, and law enforcement officials. In order to collaborate effectively, they soon found the need for a shared framework of understanding, a model they all could recognize and comprehend. They also wanted the model to create a structure for developing intervention strategies that could lead to real changes. The goal was to create an environment that would promote deep and honest dialogue, where increased understanding among a wide range of people would lead to collaboration for reducing youth violence in the neighborhoods.

Early in the project, the committee members generated a number of causal loop diagrams to depict and create a common understanding of the problem. In order to generate models that are accurate enough the project members had to rely on the wisdom and experience of those community members that were most affected by the youth violence, including the youths themselves.

It was understood at the outset that increased youth violence was due to many factors. They included poverty, social inequality, unemployment, and the belonging or lack of belonging to one or more groups. These factors were interdependent and they reinforced each other, making the problem quite complex.

3.6.1 Overall Project Design

The project was conceived as an iterative process, where four phases were identified in the work plan. There each successive phase would bring a deeper understanding of the dynamics of the system (Table 3.1).

The first phase laid the groundwork, where the project team conducted a review of academic literature and established a project plan. From this phase came a basic model. During the second phase, the project team studied one violence prone neighborhood in Boston and convened a design group made up of community workers, residents, and youth. They also assembled a panel of psychologists to learn from their expertise and thus enrich the model.

Table 3.1 The four phases of the Youth Violence Systems project

Project phase	Description
1	Review available literature Establish project plan Generate a basic model
2	Study one high-violence neighborhood Convene a design group of community workers, residents, youths, and psychology specialists Improve the model
3	Study two more high-violence neighborhoods Arrange listening sessions with gang members and their leaders Improve the model further
4	Train youth workers, community leaders, and others to use the model

There were two parts to model building: (1) creating causal loop diagrams; (2) generating interactive models using a modeling software package

The second part is discussed in the chapter on modeling and simulation (Chap. 5)

During the third phase, the project team expanded this process to include two more high-violence neighborhoods, and conducted listening sessions with gang members and their leaders. Through each successive phase, the model was improved to reflect the input of all these contributors. The fourth phase emphasized training youth workers, community leaders, youth, and many others on how to use the model to create effective conversations on reducing youth violence in the communities.

The team members continued to revise and improve the model based on what they learned, so that it could enable a variety of interested parties to understand how specific actions could affect the system and what the likely results would be. They did not expect to get a complete solution to youth violence for any given neighborhood, but they created an effective tool that could be modified to suit the needs of individual areas.

3.6.2 The Development of Causal Loop Diagrams

In order to get a better handle on the problem, the project team members generated a number of loop diagrams depicting various aspects of youth violence. These diagrams were developed in parallel with a single comprehensive diagram.

Community Trauma

The team found that many high-violence communities tended to exhibit behaviors that look a lot like the symptoms of post-traumatic stress disorder (PTSD), where individuals display high levels of anxiety and alertness. One of the consequences of the violence-induced trauma was that individuals and groups in the community could misread social cues, where a benign interaction could be construed as a precursor for impending violence. The loop diagram (Fig. 3.6) shows that with all else being equal, higher levels of community trauma led to a greater probability of high-risk interactions within the community, which in turn resulted in increased levels of community violence.

Affinity for Violence

The concept depicted here is that violence begets violence (Fig. 3.7). The loop illustrates that the more one engages in violence, the more it becomes the "default" operating mode for interaction with others. This, in turn, further fuels the cycle of violent activities.

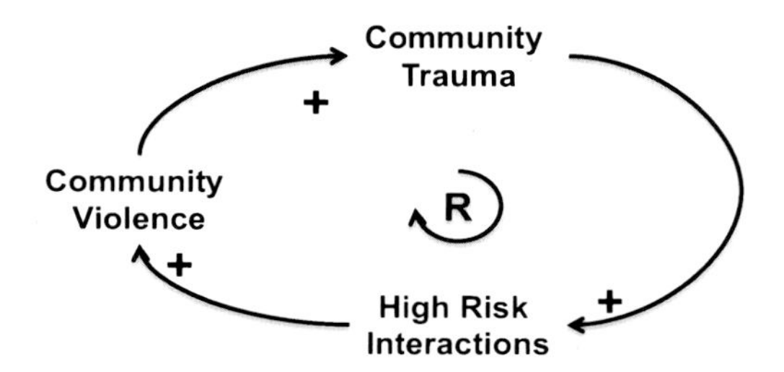

Fig. 3.6 Increasing community trauma

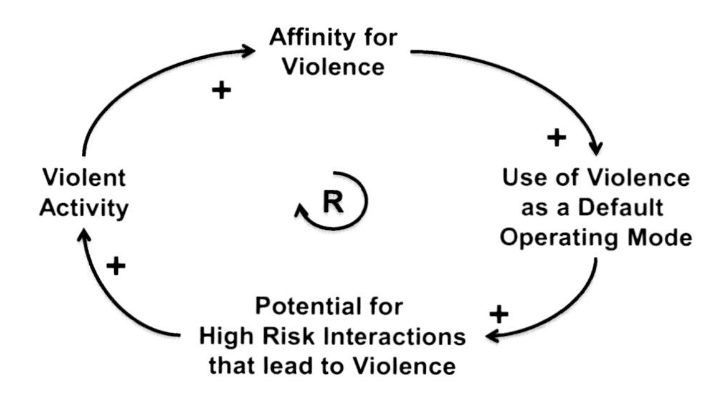

Fig. 3.7 Violence begets violence

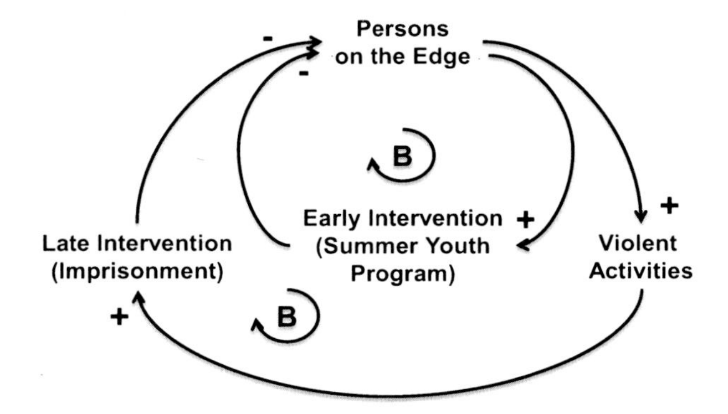

The two balancing loops that may reduce the number of persons on the edge (about to commit crime). One by early and other by late intervention. Early intervention reduces the total violent activities

Fig. 3.8 Results of early and late interventions

Early and Late Interventions

Early intervention like offering summer youth programs, which include recreational activities and temporary jobs might reduce the number of youths on the edge, who are about to commit crime. Late intervention like imprisonment after the crime could also reduce the number of youth on the edge, but would have lesser effect (Fig. 3.8).

3.6.3 A Comprehensive Loop Diagram for Youth Violence

The committee then developed a comprehensive loop diagram (Fig. 3.9 and Table 3.2), showing how a youth initially uninvolved may slide into various types of criminal activities. It showed the various levels of involvement, from being uninvolved to being directly involved including taking a leadership role. The arrows represented the pathways by which an individual may move from one category to another. For example, an uninvolved youth may become an associate, when he or she develops a relationship with someone who is involved in gang violence. The associate may then move to the edge and then commit acts of violence either as a loner or as part of a less or a more organized group.

In earlier diagrams, all individuals committing crime were put in a single group. By discussing with the community members, the committee learned more about the differences between the Loners, Less-Organized Gangs, and More-Organized Gangs (often organized as a small business). Similarly,

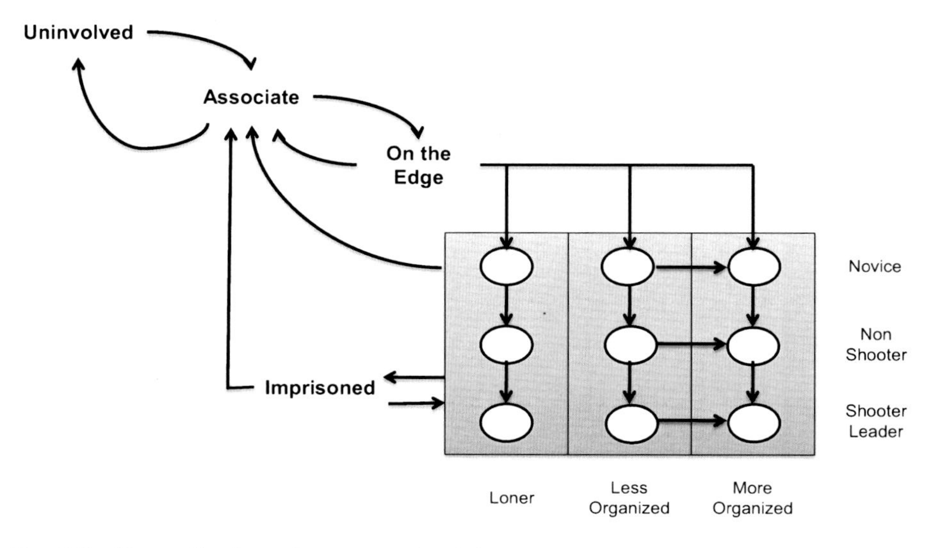

Fig. 3.9 The paths from innocence to violence

Table 3.2 Descriptions of the terms used

Term	Description
Uninvolved	One who is not committing any violent act and is not connected to a violent person or a group
Associate	One who is not committing violent acts but has close relationships with a violent person or a group
On the edge	One who has close relationships with a violent person or a group and is about to commit violent acts
Loner	One who is involved in violent acts but is not connected to a violent group
Less organized	One who belongs to a violent group that does not have a highly structured leadership
More organized	One who belongs to a violent group that is well organized and has a highly structured leadership (organized crime group)
Novice	One who is new to criminal activities
Non-shooter	One who is involved in non-gun-related violent activities like assault and battery
Shooter-leader	One who is directly involved in gun-related violent activities including leading a violent group
Imprisoned	One who is in a prison, or in a detention program and is thus restricted from directly committing violent acts

they learned about the development path for a gang member from Novice to Non-Shooter and then to a Shooter/Leader. Additionally, the diagram helped individuals, agencies, and other interested parties to identify the targets for various initiatives. A summer youth jobs program, for example, could be an effective way of moving youth back from "On the Edge" status. The comprehensive loop diagram made it easier for everyone to understand how a youth may drift towards violent activities and this led to the development of an interactive computer model, which is discussed later in the chapter on modeling and simulation (Chap. 5).

3.7 The Advantages and Limitations of Causal Loop Diagrams

A causal loop diagram conceptualizes a system by depicting its major variables and their interactions, thus, illustrating its feedback structure quite well. It follows simple rules, which makes it easy to understand and design, making system science accessible to a wide range of people.

However, the very simplicity of causal loop diagrams leads to a number of limitations. While showing the connections between variables or subsystems, it does not show their numerical relationships. The relation between the number of sales personnel and the automobile sales volume, for example, is not explicit (Fig. 3.2). It also does not specify whether the relationship is linear or not; that is, if it needs X number of sales people to sell Y number of vehicles, that proportion will still hold when the sales volume goes up or down significantly.

In Chap. 1, we discussed a system of predators and prey in a forest. There the number of wolves and deer oscillated between too few wolves and many deer to many wolves and too few deer. A causal loop diagram can show the relationship between the two species but cannot predict their proportion at any given time.

Even with all these limitations, causal loop diagraming is still widely used in depicting the relationships between different variables or subsystems in a system. This is because it can communicate quickly and concisely the essential components and their interactions in a system. In the chapter on modeling and simulation (Chap. 5), we will discuss the ways of overcoming these limitations by simulating systems with commonly available software packages running on personal computers.

Appendix III shows a set of common system behaviors that are illustrated by causal loop diagrams.

3.8 Systems Are Everywhere

It should now be quite apparent that we are dealing with systems everyday in our lives: in our home, at work, or in recreational activities. Thus, it is always better not to look at an object or an event in isolation but to perceive it as a part of a closed loop system. While most ancient cultures gave due emphasis to the cyclic and interconnected nature of all worldly phenomena, we in the modern world often tend to forget that, to our own detriment.

Systems Approach in Our Daily Lives

Back in my younger years with two teenage daughters, my wife and I found ourselves in a constant battle. The teens enjoyed shopping and seemed to want all the latest trends. "All the other girls at school have these jeans," they complained, "Why can't we?" Even though our eyes were popping at the price tags, no amount of rationalizing could convey the impracticality of buying two pair of designer jeans for the amount we might have outfitted them from head to toe with the more common brands. Thus, every shopping trip with them led to arguments, grumpy faces, and tears.

My wife and I, tired of arguing with the girls and sounding like stingy parents, finally handed them each the monthly amount we budgeted for their expenses and told them to figure out how to clothe themselves within those limits. We provided them with one caution—too many fashionable clothes in the spring might mean no winter coat in December.

We braced ourselves for the reckless spending they had tried to push on us but we were shocked. They started to read price tags, cut coupons, and look for sale items. They balked at outrageous prices, weighed options, and strategized about making the most of each dollar. Apparently, we had raised two market savvy daughters. They also were eager to meet the challenge of staying under budget to prove that they could do it better than us. Designer jeans never showed up in the house, and in fact, they did a good job of meeting their clothing needs and still had some money left over to put in the bank.

You might say what we gave them was independence; however, it was more than that. What we gave them is the ownership of the process, allowing them to understand all the feedback loops. Before this experiment, they had desires and fulfillment or in many cases non-fulfillment. Once they had control of the money, they still had desires but those were tempered by factors like usefulness, price, and value. We had to admit, they did a great job. They seemed to be happy with a few trendy items. Additionally, they were not left wearing fancy clothes without warm coats through the long New England winters.

We were happy not to have to argue with our daughters or manage their budgets and they were glad to be in control. What it took was the creation of a scenario where a complete system resided with them, rather than the previous situation where half the system was with us and the other half with them making it more like an open system.

Similar problems exist in organizations, where managers try to micromanage their subordinates. Ideally, higher-level managers should set the strategic visions and agree on objectives and targets while leaving the operational decision-making and execution to their subordinates.

Politicians Can Better Serve Their Constituents by Taking a System Dynamic Approach

Professor Forrester made the apt remark that most of the pressing problems facing human societies today cannot be addressed properly until a new generation of leaders familiar with system dynamics takes leadership roles. Politicians often make decisions based on a hunch or gut feelings that have far-reaching consequences to a society or a nation. If a decision produces the desired results, the politician may get due adulation from his or her contemporaries and later from historians. On the other hand, if a decision produces adverse results, then the politician may lose the next election and may face vilification from future historians. By doing due diligence, using a system dynamic approach, a politician may be able to make more informed decisions than would otherwise be possible.

Politicians and military strategists might also use the system dynamic approach to better analyze the motivations of adversarial nations and their leaders. Politicians may also use this approach to make long- and short-term economic decisions. There is almost no area in the political decision-making process that could not benefit from the system dynamic approach.

4

The Fundamental Behavior Patterns

Abstract A critical step in understanding the behavior of systems is to be able to identify their key patterns. A system may be in a steady or a dynamic state, but steady state is more pervasive, where the various forces acting on a system are in balance thus producing little change. Dynamic behavior, on the other hand, is caused by changes in these forces that lead to imbalance, which eventually leads to an altered balanced (steady) state. The fundamental patterns of dynamic behavior are growth, decay, oscillation, and goal seeking. Oscillation is often caused by high gain and delays in a system. The interaction between a courting couple, for example, is often a high gain system, where minor disagreements lead to volatile situations.

This chapter starts with a discussion on steady and dynamic behavior patterns, along with nonlinearity and time variance. Various examples of growth and decay are presented, which include the compounding growth of a bank deposit, the Chernobyl disaster, and China's rapid economic growth. These are followed by discussion on oscillation, instability, and goal seeking. Finally, the difference between detail complexity and dynamic complexity are examined along with the behavioral unpredictability of some systems.

4.1 The Steady State Behavior

When a rock is resting on a hillside, its position with respect to its neighborhood is stationary. It is held in position by the interactions of various forces, such as gravity, which is acting to pull it down, and frictional forces and inertia, which inhibit it from doing so. In addition, the soil underneath the rock, which is

A. Ghosh, *Dynamic Systems for Everyone*,
DOI 10.1007/978-3-319-43943-3_4

undergoing stress due to its weight, generates a restoring force that keeps it from sinking into the ground. Thus, the rock is in a steady state. Similarly, a parked car is in a steady state. However, steady state does not always mean inactivity. When the car is being driven at a constant speed, say at 60 miles an hour, that may still be considered to be in a steady state (Fig. 4.1). The car may also be considered to be in a steady state when it is moving with the flow of traffic keeping its position nearly constant with adjacent vehicles. However, if the position of the car with respect to its geographical location is of primary consideration, then a moving car will be considered to be in a dynamic or non-steady state.

When I am sitting quietly after my dinner, I am in a steady state, even though my stomach is working feverously to digest the food that I have just eaten, my lungs are working to provide oxygen to my blood stream, and my heart is pumping blood to my various organs. A system is in steady state when the variables that are under consideration continue to remain the same as they were in recent past and will be so in the immediate future. Thus, the state of system whether steady or dynamic often depends on the perspective from which it is considered.

A system is generally in a steady state a majority of the time. That is because the various forces that are acting on it tend to influence the system toward a stable condition. For a living system, the word homeostasis is often used to describe the state of balance, which is maintained through a series of feedback

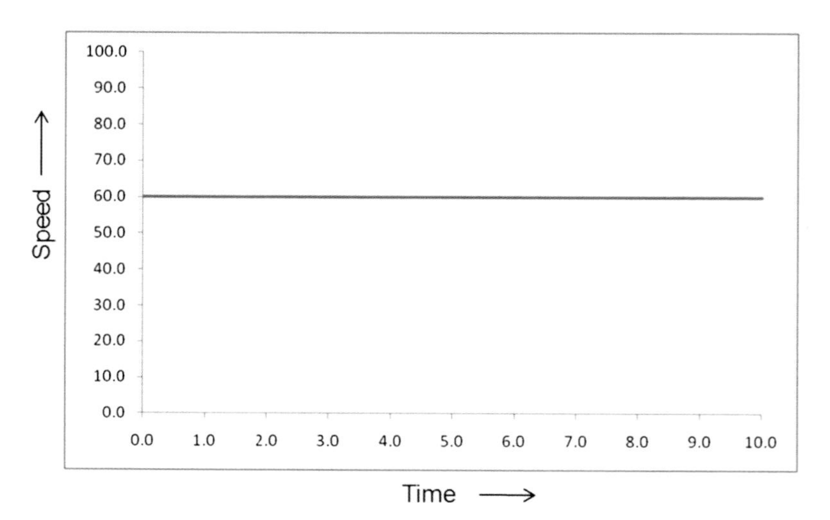

Fig. 4.1 Steady state behavior

mechanisms. The Chinese concept of yin-yang specifies how complementary opposite forces interact in a dynamic system, biological or otherwise, to give rise to a balanced state. However, a steady state may not be the same as the most desired state. For example, the highway system in a town may be in a steady state of disrepair, or the achievement scores of students in a school may stay steadily below expectations.

In many systems, steady state is not achieved until some time has elapsed after the system is started or initiated. Similarly, any major change in a system may lead to an unsteady condition, which may eventually return to a steady state. This unsteady condition is often called as a transient state.

4.2 Nonlinearity and Time Variance

Before we delve into dynamic behavior patterns, let us look at two important issues that contribute to the complexity in system behaviors. They are nonlinearity and time variance.

4.2.1 Linear and Nonlinear Behaviors

A system behavior pattern is said to be linear if the ratio of change of an output due to a change in its input is constant. In other words, the output varies in a straight line for the variation of its input (Fig. 4.2). For example,

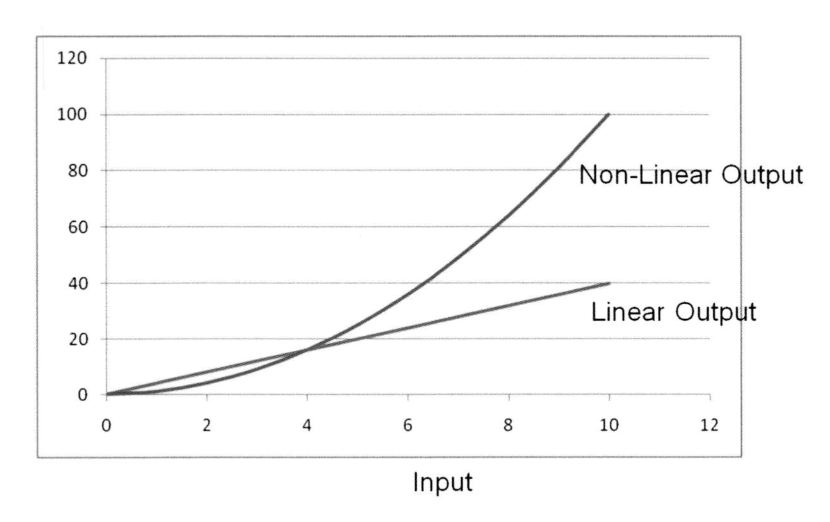

Fig. 4.2 Linear and nonlinear behaviors

if I sell one pair of shoes and make a profit of $3 and then I sell five pairs of shoes and make a profit of $15, then the relationship between the profit and the number of shoes sold is linear. Similarly, when I open the tap, one turn and get a flow of two gallons of water per minute and when I open the tap two turns I get four gallons of water per minute, then there is a linear relationship between the input and the output.

However, if I open the tap one turn and get two gallons of water per minute and then I open another turn and I get six gallons of water per minute, then the relationship between the input and the output is not linear. Nonlinear systems are more common in the nature, such as the growth of a human baby with its age or the attraction of a magnet to a piece of iron with the distance between them.

In the case of selling shoes, the profit may be linear with the number of shoes sold for a small variation of numbers. However, if I start selling shoes in very large numbers, then my overhead cost per pair of shoes will decrease and my net profit per shoe sold will increase, making the relationship nonlinear. Optimization of nonlinear systems is inherently more difficult than linear systems; however, in many practical cases a system may be considered linear in a relatively small operating area.

4.2.2 Time Invariant and Variant Behaviors

A system output is time invariant if it produces the same response to a fixed input regardless of the time. If I open a tap one turn and get a flow of two gallons of water per minute, whether it is in the morning, or in the afternoon, or in the night, then the flow rate is time invariant. However, it will be time variant if the flow varies due to the variation water pressure over that period. Most biological and social systems are time variant, while mechanical systems are less so. The mood of my boss may vary from one day to another, which has to be taken into account when dealing with her. The reaction of a child to a new toy today may not be the same a year later, when she has grown a year older. Even inanimate things change with time. If I go back to the town where I spent my childhood, I may find it difficult to navigate there after my absence of two decades, because many of its neighborhoods and some of its road systems have changed. My mental models have to be altered to adjust to the new situation.

4.3 Dynamic Behavior Patterns

A system does not stay in a steady state all the time because it is usually subjected to varying external and internal events and stimuli. In addition, the actions caused by a system can alter its environment and thus change the forces that in the past kept it in a stable state.

A hot water supply system, which was described in Chap. 2, heats cold water to the required temperature using steam (Fig. 4.3). A sensor measures the hot water temperature at the outlet and the controller maintains the required temperature by manipulating the valve at the steam inlet. Let us assume that the system is in a steady state, that is, the water at the outlet is constant at the required temperature. However, there are a number of factors that can upset this steady state, which include change in the demand for hot water, change in cold-water temperature at the inlet, and the temperature of the heating steam. Additionally, there may be a need to alter the desired temperature of the hot water. Any such change will of course lead to corrective actions by the controller but there will be instability, preferably for a short period, before the system settles down again to an altered or changed steady state.

A manufacturing facility is in a balanced state when its manufacturing capacity is equal to the demand for its widgets (Fig. 4.4). Then the gap between the demand and the production rate is zero. The gap will increase if the demand for the widgets increases suddenly beyond its maximum manufacturing capacity. This will prompt the management to increase manufacturing capacity, but that will take time. Therefore, the gap will persist for some time until it reaches zero, leading to another steady state.

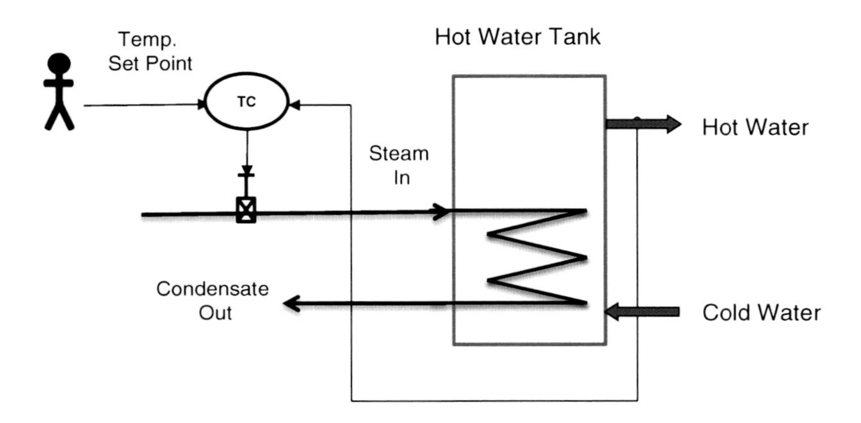

Fig. 4.3 Hot water temperature control

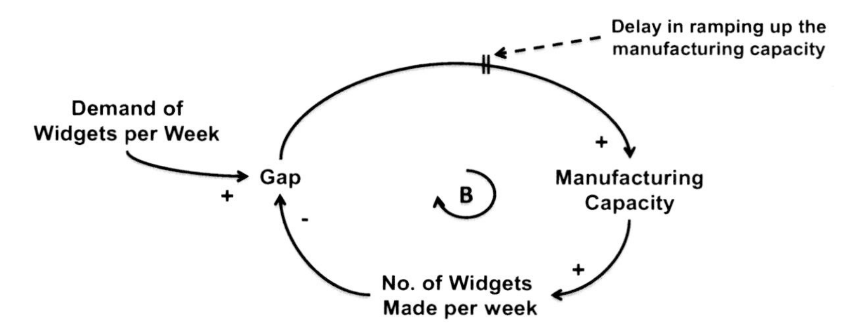

Fig. 4.4 Changing demands in a manufacturing facility

4.3.1 Growth and Decay

There are many reasons for the growth and decay of a system or its outputs. They include various external causes, such as abundance or shortage of stimuli or resources, natural decay (old age), or a change of its objective or goal. The growth of a tree may be stunted because of prolonged drought. Rat population may multiply because of the greater availability of grain lying on a field. The economy of a nation may grow rapidly with the loosening of trade barriers. A driver of an automobile can decide to change its speed by increasing or decreasing the pressure on the gas pedal or a person may alter the temperature of a room by changing the set point of the thermostat.

4.3.1.1 Linear Growth and Decay

Linear growth or decay follows straight lines, which are not that common in real life (Fig. 4.5). An example of linear growth would be the change in the immigrant population in a country, where the country allows only a fixed number of persons to immigrate per year. If a person saves a fixed amount of money every month and puts it in her home piggy bank, then her savings will follow a linear growth path. If I am driving a car at a steady speed on a flat terrain or on a hilly road with constant slope, then the gasoline in the tank of the car will be depleted at a steady rate; that is, it will exhibit a linear decay.

4.3.1.2 Exponential Growth and Decay

Exponential growth or decay occurs when the amount of growth or the decay of a variable is proportional to its current value. For example, the growth of population of a country in a given year depends largely on the actual

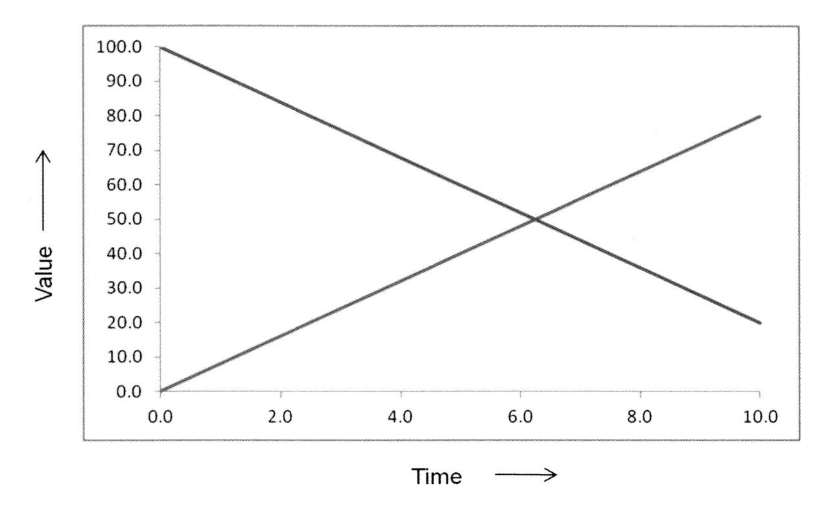

Fig. 4.5 Linear growth and decay

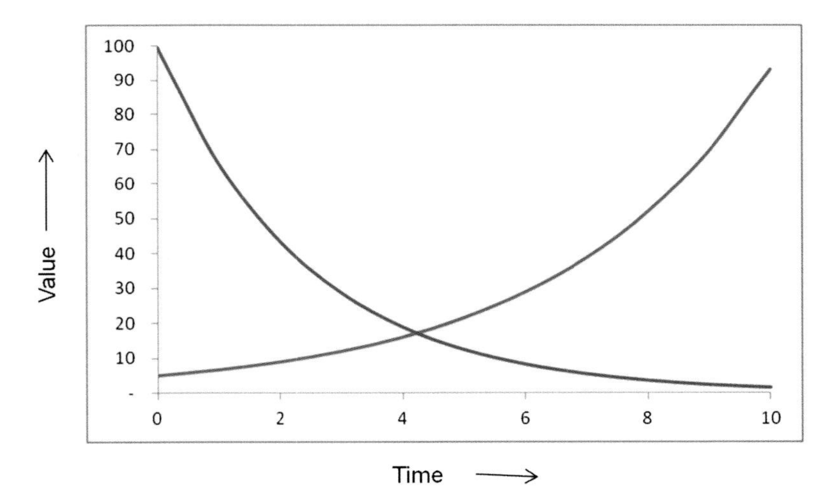

Fig. 4.6 Exponential growth and decay

population of that country in that year. The greater the population the greater is the net growth. In that scenario, even when the rate of growth remains constant the actual growth goes up every year (Fig. 4.6). Exponential growth and decay is much more widespread in nature than the linear ones.

4.3.2 Sally's Bank Account

Today is Sally's 65th birthday and she is feeling very happy as she is retiring from her work. She is planning to take a long cruise vacation with her husband Eric. A few days back she received a letter from her bank about a certificate of deposit (CD) in her name that had matured with nearly $60,000 in the account. That was a pleasant surprise, but it made her wonder how she got that much from account about which she had little recollection. Then she remembered that on her fifth birthday her doting grandparents bought her a $1000 CD as a birthday present (Table 4.1). Sally's grandparents, who were long deceased, had remarkable foresight. The fixed-term 60-year CD they bought for her for $1000 offered 7 % interest per annum. It has now grown to nearly $60,000. How could that happen? This shows the power of compounding (or exponential growth), where the amount of growth of a fund in a given year is proportional to its current value. The CD grew by $967 in the first 10 years, but in the next 10 years, it grew by $1913 and $3742 in the following decade and so on, even though the interest rate of the bank account remained constant.

If a bank account with 7 % interest rate can grow so much over time, how much can a bacteria colony with a much higher reproduction rate grow in a much shorter period? Let us say that a bacteria colony doubles itself in 24 h. Then the small colony of 1000 bacteria will grow to 32,000 in a mere 5-day period and over a million in 10 days, provided of course the conditions for the growth continue to be the same during that period (Fig. 4.7).

4.3.3 Rapid Escalation of Arguments Between a Courting Couple

Julie and Ron are passionately in love. They are engaged, and planning for their wedding. Julie lives with her mother, while Ron lives in an apartment with his mates in another part of the town. They both like outdoor sports like

Table 4.1 Exponential growth of the bank deposit amount

Year	Bank deposit amount (US$)	Increase in a decade (US$)
0	1000	
10	1967	967
20	3870	1903
30	7612	3742
40	14,974	7362
50	29,457	14,483
60	57,946	28,489

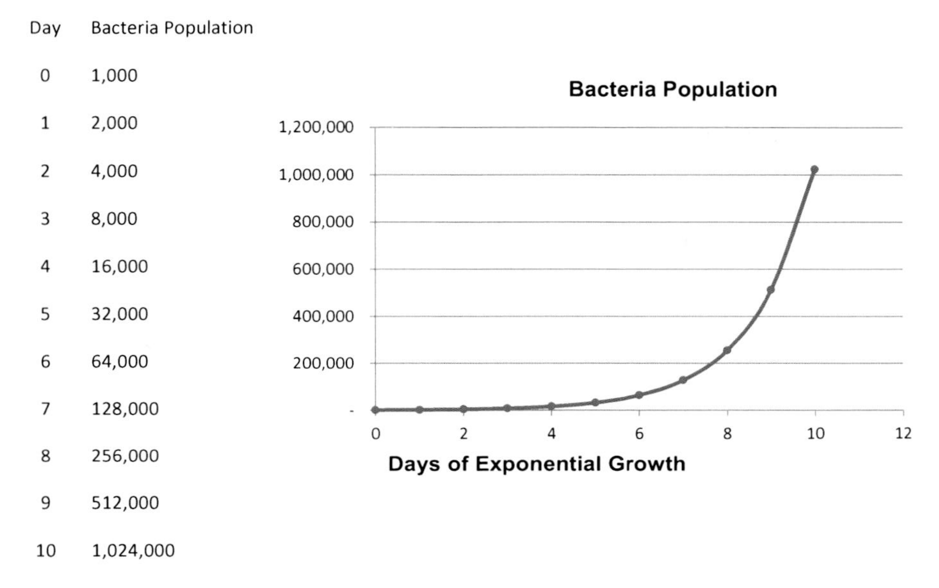

Day	Bacteria Population
0	1,000
1	2,000
2	4,000
3	8,000
4	16,000
5	32,000
6	64,000
7	128,000
8	256,000
9	512,000
10	1,024,000

Fig. 4.7 Exponential growth of a bacteria colony

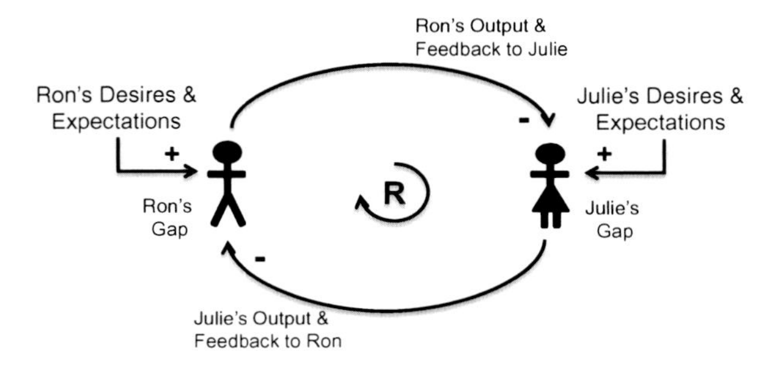

Fig. 4.8 Ron and Julie's interactions

hiking and skiing but also enjoy going to cultural events. Lately, they have become inseparable. In their spare time, they are either together or talking to each other on their phones.

It was a Sunday morning in the early spring, and both were thinking of how to spend the day. Ron called Julie and the conversation went like this (Fig. 4.8):

Ron: Hi honey, I've been thinking about you since I woke up. I went jogging and now I'm having my morning coffee with an oatmeal–cranberry muffin.

Julie: Hi love, I'm having my breakfast too and I've been thinking about you in my dreams.

Ron: Do you have a plan for the day? The weather's nice and I found out there's still some fresh snow up in the White Mountains. How about we go there to ski today? This may be our last chance this season.

Julie: You know the famous Kirov Ballet is in town and they're dancing Swan Lake this evening, I'd really like to go there.

Ron: We've been to the ballet before. This is possibly our last chance to ski this season and the weather's ideal.

Julie: I've skied with you a lot and don't want to miss Swan Lake, especially the Kirov Ballet. If you don't like what I want then why not go to White Mountains with your mates?

Ron: Julie, you know I don't like doing anything without you.

Julie: Then why are you hurting my feelings? I've always agreed to what you wanted. Now, you're bullying me about going skiing.

Ron: Am I a bully? If that's what you think about me, very well...

Ron slams the telephone down and that ends the conversation. For a fleeting moment, Julie thinks about breaking off the engagement. However, the story has a less tragic ending. Later in the day, when their emotions have calmed down, Julie calls Ron. It was too late to go skiing and all the tickets for the ballet were sold out. So, they decided to spend the evening at a movie.

This is an example of reinforcing loops with high gain, where signal amplifications lead to an escalating situation. Both Ron and Julie are in love and are deeply attracted to each other. The difference between Julie's desires and expectations and the feedback she was receiving from Ron was highly amplified and the same was the case with Ron. That led to a rapid escalation of their perceived differences. However, when their tempers cooled, the signal amplifications between them were reduced, leading to a more manageable situation. They could have had the situation under control earlier if one of them had stayed calm during their conversation thus reducing the gain in the system.

4.3.4 The Chernobyl Disaster

The accident in Chernobyl nuclear power station in the Ukraine occurred on April 26, 1986, at 1:23 AM during a test of a turbine generator on the Unit 4 reactor. It happened when control rods were improperly withdrawn while an important safety system was inactive, in violation of the operating rules (Fig. 4.9). That caused the reactor to overheat, explode, and catch fire. As the

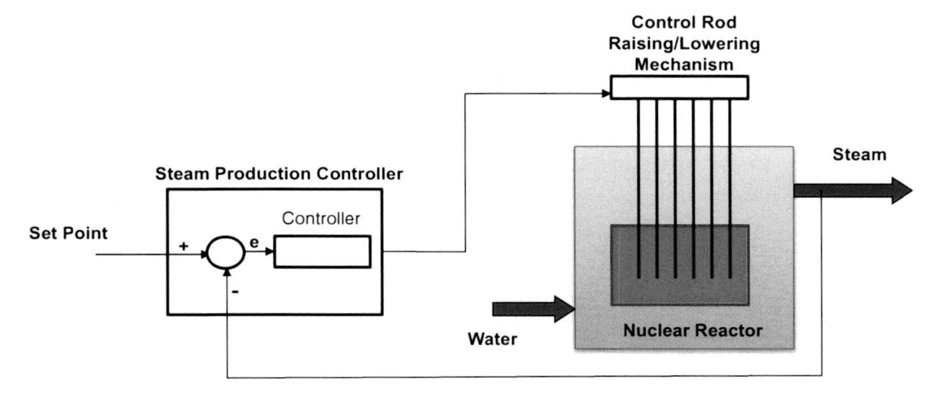

Fig. 4.9 Nuclear reaction control

facility lacked an adequate containment structure, the damage to the reactor core and the control building allowed large quantities of radioactive material from the reactor core to be released during the ensuing 10 days. This necessitated the evacuation of tens of thousands of people and farm animals from the surrounding area. The release of radioactive material to the atmosphere resulted in radiation sickness and burns to more than 200 emergency personnel and firefighters, and over 30 fatalities.

A nuclear chain reaction takes place when radioactive materials such as uranium or plutonium are bombarded with sufficient neutron particles. These neutron particles are generated by the radioactive materials, so a chain reaction takes place when there is a critical mass of radioactive material as in the case of a nuclear bomb. The control rods play a critical role in controlling the reaction rate in a power reactor by absorbing some of the neutron particles. Therefore, the control rods are raised or lowered to control the steam production. The rods are lowered completely into the uranium bundle to shut the reactor down in the case of an accident or to recharge the fuel. Thus, in normal operating conditions the nuclear reaction is kept in a steady state by adjusting the position of the control rods.

The nuclear chain reaction is a typical example of exponential function. Other examples include exothermic chemical reaction, growth of bacteria, growth of cancerous cells, growth of bank balance with compounded interest, and the growth of membership in a successful pyramid or Ponzi scheme. Often, exponential growth is quite rapid and happens mainly when there is large positive feedback and little or no negative feedback.

In real life situations, exponential growth cannot be sustained for long periods. The effects of that growth can also lead to its demise, as happened at Chernobyl. A forest fire started after a long drought usually spreads rapidly burning up all the nearby trees and other vegetation and then ends naturally even without human intervention. Similarly, when a pandemic spreads in an exponential fashion it is contained naturally with the death of a large number of susceptible persons and by increasing the natural immunity among others. However, in many situations we cannot wait for the natural turn of events, because the cost of inaction can be too high. An uncontrolled forest fire can devour many human habitats and a pandemic can kill many people and wreak havoc on an economy. To minimize the damage we need to intervene long before such an event ends naturally.

4.3.5 China's Economic Growth

Looking at economic growth, everyone's attention now turns to China. For a number of years, China has sustained an economic growth rate of around 10 % per annum. That means China's economy has been doubling every 7 years. This remarkable achievement has lifted millions of Chinese households from poverty. However, from a system's perspective, this rate of growth cannot be sustained any more due to a number of negative feedbacks that are starting to affect its economy (Fig. 4.10).

The main factors that led to the growth were:

- Reforms of 1978 that led to privatization of many businesses and industries
- Low labor cost
- Low interest rate
- Undervalued currency
- Large overseas investments
- High demand of Chinese goods overseas

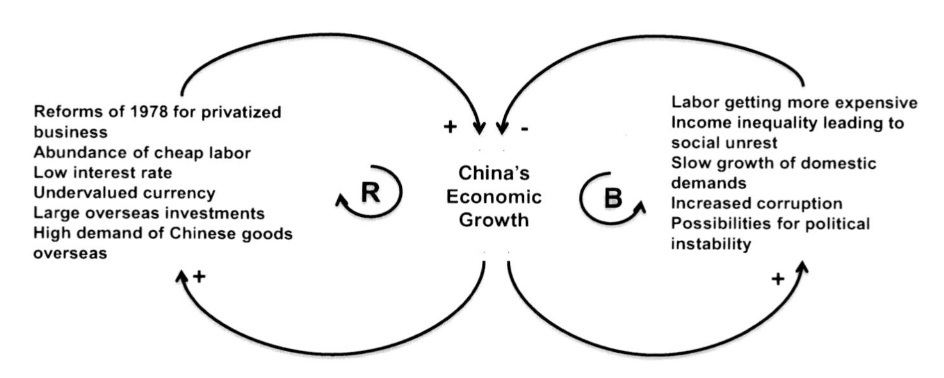

Fig. 4.10 The economic growth of China

The factors that are starting to inhibit the growth include:

- Labor is getting more expensive
- Weak innovation capability of Chinese industries compared to Western counterparts
- Income inequality leading to social unrest
- Slow growth of domestic demands
- Possibility of increased corruption
- Possibilities for political instability

Without doubt, the era of cheap labor is ending, due to the economic growth, which has sucked up a large part of the rural population in many areas. This is leading manufacturers to move their production facilities inland where wages are lower. Additionally, the one-child policy has created a demographic disadvantage. Smaller number of working-age persons will soon have to support an increasingly larger number of seniors. In addition, there will soon be a saturation of overseas market for Chinese products, which will induce the authorities to encourage increasing domestic consumption. However, that may not be enough to sustain the high growth rate of the yester-years. Lower growth rate with increased income inequality will exacerbate social and political unrest.

In order to sustain economic growth China has to be more innovative. That will require more proactive actions on the part of the government, which include:

- More freedom for expressing new ideas that may not be in line with current policies
- More investment in research and development by private companies
- Active encouragement from the government by offering tax incentives for private research and development
- Rigorous enforcement of intellectual property laws
- Better application of antitrust laws to foster competition
- Allowing private investors to risk money on ideas that may or may not work

After all, China has a long history of innovation, which includes compass, gunpowder, paper, and the printing press. With the right environment, China can again become a leader in developing new ideas and products.

4.4 Oscillation and Instability

Everyone is now familiar with the term Economic Instability. National and world economies go through cycles, which are periodic in nature. Russian economist Nikolai Kondratiev was one of the first to propose the theory that Western capitalist economies have 50–60 year cycles of boom and bust. These business cycles are called "Kondratiev waves." However, a number of other economic cycles both major and minor are in play. Major cycles last longer and are less frequent, while minor ones come and go more often. They superimpose on one another to produce perpetual economic turmoil and uncertainty.

A driver driving a car on a highway with light traffic is generally able to maintain a steady speed by keeping an even pressure on the accelerator. Suppose, the car moves from a lower to a higher speed limit area. The driver may then choose to increase the speed by depressing the gas pedal down a bit further. A good and experienced driver will intuitively put adequate pressure to accomplish this task. However, if the driver is inexperienced or has a problem with her foot that makes her press the pedal too hard the car will run at a faster speed than she intended. On finding the car going too fast, the driver may decide to ease off the pressure on the gas pedal and may over correct leading to a slower than intended speed. The driver may then push the gas pedal again leading to a higher than intended speed and the process will be repeated. This is an example of oscillation, where the controlled variable goes up and down periodically. The reason is over correction, which often happens in a high gain system.

Oscillation is also common in a system with large lags. At an automobile dealership, for example, the inventory of cars at any given time is the result of sales volume and the delivery of new cars from the manufacturer. The sales are happening all the time and orders for new cars are sent as soon as inventory falls below a threshold. However, the delivery from the manufacturer usually gets delayed causing variations in the inventory level (Fig. 4.11). It is found that the larger the delay the larger is the variation of the inventory level. Delays in the supply of new vehicles beyond a threshold lead to oscillation of the inventory level at the dealership.

A system is prone to oscillation, where there is a delay in the system (which is common in most systems) and the gain is high (Figs. 4.12 and 4.13). Thus if too much stimulus is given to a sluggish economy, then that increases the possibilities of oscillation and instability (bubble formation). The politicians and economists face the dilemma where too little stimulus may make the economy grow at an unacceptably slow rate, whereas too much of it can lead to oscillation and instability.

When the gain is low, the system may show sluggish response; with increased gain, the system may start oscillating. If the gain is not too high, the

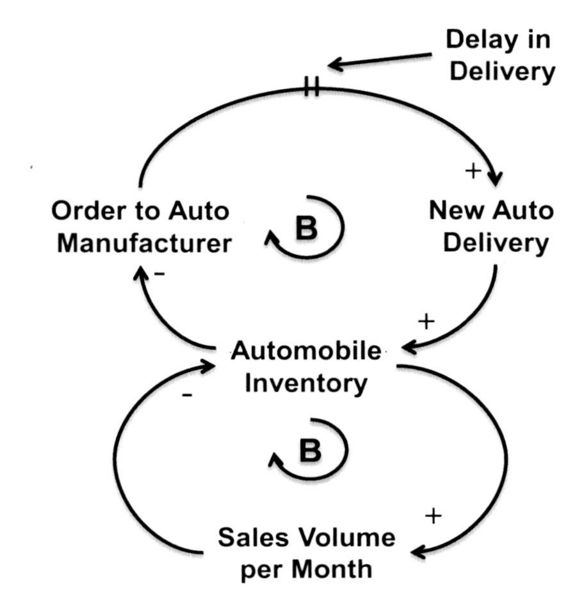

Fig. 4.11 Oscillation in automobile inventory level

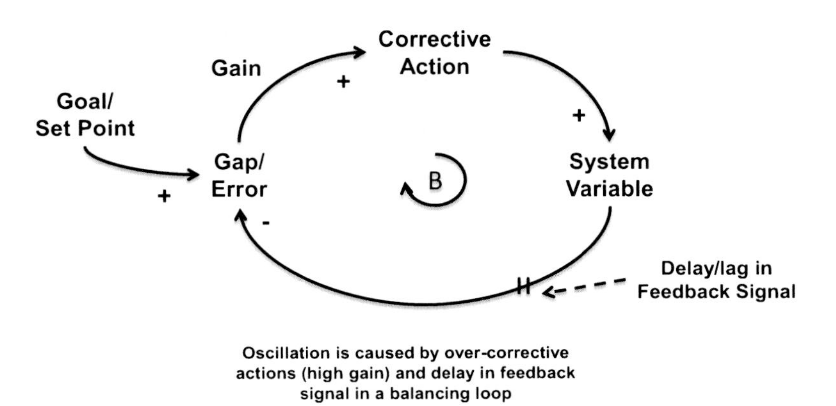

Fig. 4.12 A typical oscillatory system

oscillation may be of a decaying type, where its amplitude gets less and less with time, until it is stabilized. This temporary oscillation may be acceptable in many situations, such as for an economy, but less acceptable in other places like an airplane during a flight.

When the gain gets even higher, a system may show a sustained oscillation or one with progressively increasing amplitude. In an industrial system, a sustained oscillation may quickly wear the parts in a machine. However, sustained oscillation may be acceptable in certain situations but oscillation

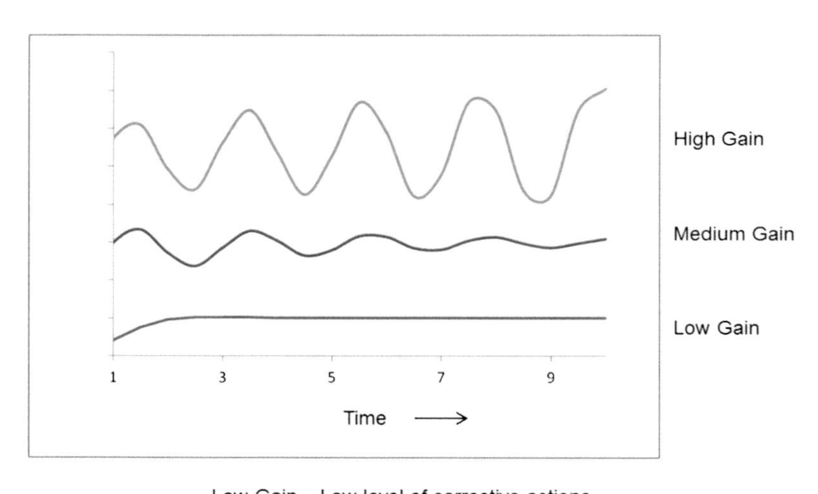

Fig. 4.13 Oscillations

with increasing amplitude is rarely tolerated. Increased oscillation can cause output to exceed its safety limits leading to hazardous conditions causing damages to the system and its environments. One common course of action when a system oscillates is to reduce the gain in the system, if possible. Otherwise, shutting the system off may be the course of action. In the case of an airplane in flight, the best course may be to land on the nearest possible runway, while for a nuclear reactor it may be to shut it down immediately.

Another point for consideration is the cycle time of oscillation, which is the elapsed time between one peak and another. The cycle time is usually not affected by the gain but varies with the delay or lag in a system. The longer the delay the longer is the cycle time of oscillation. A business with a short turnaround time, measured in days between order and supply, may experience cycle time measured in days, whereas the economy of a country with all its long lags will have oscillations with a cycle time measured in years. Things get more complex in the case of the economy of a nation where there may be a number of lags of differing lengths, which give rise to a number of oscillations with differing cycle times, occurring simultaneously.

4.4.1 A Firsthand Experience of System Oscillation

I had a personal experience of violent oscillation of an engineered system in an oil refinery where I was working early in my professional life. I was looking after a plant where hydrocarbons that are normally in liquid state were extracted from natural gas, and then used in blending gasoline. The natural gas was delivered from a nearby oilfield in a 6-ft diameter pipeline. The downstream pressure of the natural gas to the plant was maintained by closed-loop control, which included a pneumatically operated giant-sized control valve. The valve was slow to move from one opening position to another because of its size and because it was operated by pneumatic pressure. A chart recorder recorded the downstream pressure. One day, when I was still a novice, I noticed that the pressure as recorded on the chart was drifting erratically around the desired set point. That led me to believe that the response of the controller was sluggish and increasing its gain would make it more responsive to the upstream pressure disturbances.

My diagnosis was correct but taking the right corrective action was not that simple, which I soon realized. I increased the controller gain a bit and found that it reduced the drifting of the downstream pressure readings. Encouraged by the result, I increased the gain a little bit more, which reduced the drifting even more but did not eliminate it altogether. So I felt confident to increase the gain even more to get a more precise control. However, as soon as I increased the gain for the third time, something very strange happened. Instead of reduced drifting and precise control the chart recorder started to show wide swings in downstream pressure. I could hear the control valve opening and closing wildly, and the 6-ft-wide pipeline was shaking as if an earthquake had struck. This was quite unexpected and frightening to me. Luckily, a more experienced colleague who was working nearby promptly came to help. He immediately reduced the controller gain and everything returned to a near normal state.

The lesson I learned is that though increasing the gain in a system may lead to better control it also carries the danger of oscillation and instability. The way out was to tune the controller properly, which included adjusting other controller variables, such as integral and derivative control actions along with the proportional gain (Appendix I).

4.5 Goal Seeking

Goal seeking or maintaining the status quo if it has reached its desired state is an essential part of a system. Without that, a living system will not be able to survive or progress and an electromechanical or industrial system will not produce products of desired quality or quantity.

When a system is trying to maintain its existing state it is likely to face various extraneous forces, which may be called *loads* that may affect its current situation. I may be feeling comfortable now but a sudden drop in the atmospheric temperature will cause me run for my warm sweater. In the mean time, my body will react by pumping more blood to my skin and other affected parts, such as fingers and toes. Similarly, a distillation column in an oil refinery needs to maintain the quality of its distillate outputs in the face of sudden change in the composition of its crude oil input.

How well a system reacts to load changes depends on its characteristics. In the case of a biological system, like my body, there are complex control systems, some at the lowest cell level, which take instinctive actions. Then there are others at the highest (brain) level, which makes conscious decisions, and some in between.

In the case of a distillation column, the situation is different. The column itself is a static tower made of steel and other materials with no intelligence of its own. It is driven by a control system, which manipulates its various process parameters, such as rate of input and output flows, their temperatures, and pressures. Thus, the control system drives the goal seeking part of a mechanical system. The two parts together define the characteristics of the distillation column.

Another type of change takes place when a system's goal is changed. If I am feeling cold, I may adjust the set point of the thermostat in the room. In the case of a distillation process, the production manager decides to alter the composition of some of the distillates. A person driving a car at steady speed of 40 miles/h the driver may decide to change its speed to 65 miles/h, when he gets on to a highway. These are all examples of goal changes.

The behavior of a system when facing disturbance due to load change or goal change is based on its characteristics and is similar for both the cases. Ideally, it should be able to act instantaneously when that occurs and correct the situation immediately, but in reality that is seldom possible. A home heating system will take a significant time to heat up a room to a new set point. An automobile will take a finite amount of time to accelerate from 40 to 65 miles/hour. An automobile with enormous power and low inertia (little weight) would be able to reduce the time for acceleration considerably but will not be able to eliminate it altogether.

When a goal or the set point of a system is changed, the relevant output of the system starts changing quickly at first but slows down as the difference between the goal and the output (error) narrows. If the system has integral control action in addition to proportional control, with a low or moderate gain, the system continues to change until it reaches the goal (Fig. 4.14). However, if there is a significant second order lag, that is there are two lags in series, then the system will follow an "S"-shaped curve (Fig. 4.15), where the two lags together will slow down the initial changes.

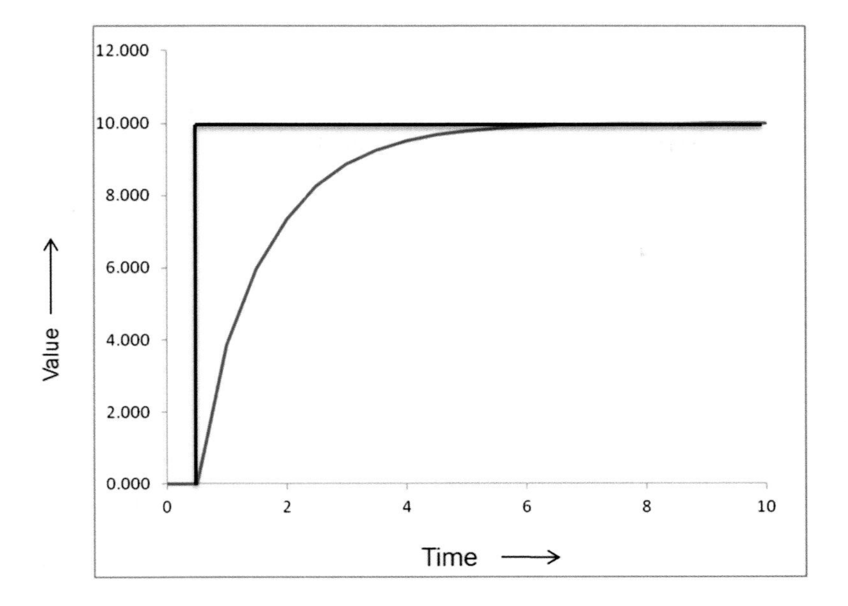

Change of output with a step change in set point

Fig. 4.14 Goal seeking

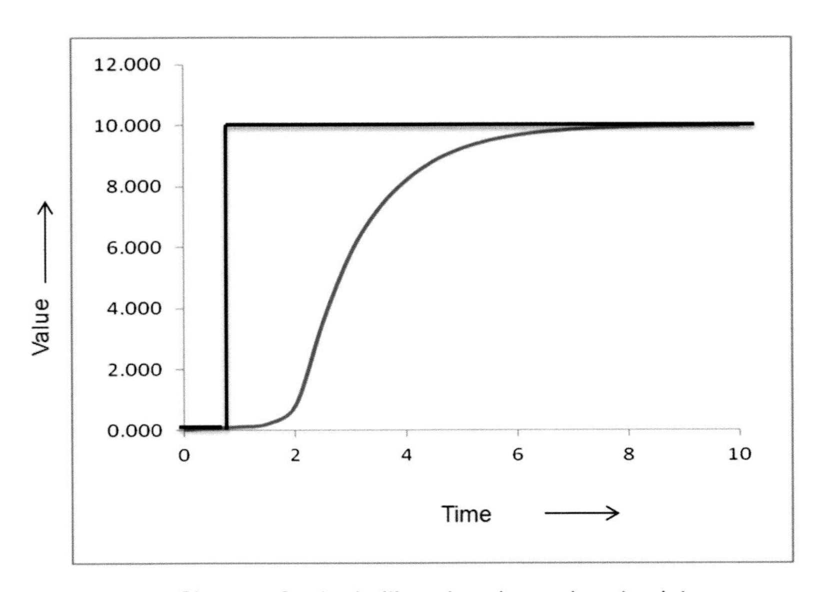

Change of output with a step change in set point

Fig. 4.15 Goal seeking with a lag

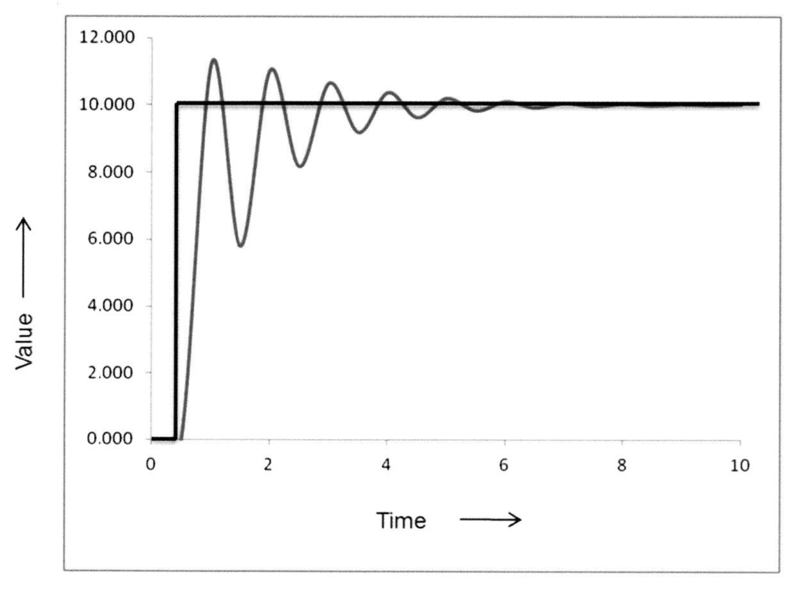

Change of output with a step change in set point

Fig. 4.16 Goal seeking with oscillation

If the gain in the system is increased, then it will reach the new goal faster but may overshoot and oscillate before settling down to the new goal (Fig. 4.16). If the gain is increased even further, then the system may go into a state of sustained oscillation or one with progressively increasing amplitude, which will likely have detrimental effects on controllability or safety of the system. The best course of action therefore is to reduce the gain of the system enough to allow a tolerable amount of decaying oscillation, without making the system too sluggish.

4.6 Detail Complexity Versus Dynamic Complexity

A large system generally has many variables that interact with each other to create a complex system. A water supply system in a large city is complex, but under normal operating conditions, it is a stable, that is it does not change very much from one day to another. Once it is set up properly, the system goes on supplying water to the residents without interruption for 24 h a day and 7 days a week. However, its dynamic complexity may become apparent in an abnormal situation, such as the rupture of a major supply line to an area in the city. The problem may be minimized by alternate routing of the water supply to that part of the town. However, that may have an untoward effect, such as reduced water pressure to another area. In this scenario, examining the dynamic behavior of the system for possible failure conditions can be helpful.

Most human relationships are dynamically complex while relatively simple in detail. The relationship between parents and a child evolves from the child's birth to her maturity and beyond. The doting parents the baby had in her earlier years are often replaced, for good reasons, by stern disciplinarians in her childhood and teen years. The sweet child of early years may change to a rebellious teenager. Then, as the child grows to maturity, the relationship often changes to one of cooperation and friendship. A relationship may go sour if either party does not realize its evolving nature and pay enough attention to the need for change.

Detail complexity involves a large number of variables but they do not change with time. However, in dynamic complexity, the cause and effects are subtle, and their effects over time are significant, which are not always apparent beforehand. In studying the behavior of a system, the appreciation of its dynamic complexity often offers greater rewards than the study of its detail complexity.

4.7 The Puzzling Behaviors of Systems

It often seems puzzling when the same system produces different sets of behaviors at different times under seemingly identical circumstances. That happens more often with nonengineered than with engineered systems. Engineered systems are generally designed to produce predictable results and are thus expected to produce consistent behavior within their normal operating ranges. However, that is rarely true for social, economic, and biological systems. The varying sets of behavior may be explained by the differences in the initial conditions and are often due to the randomness of the environment around them. In the example of predator and prey in Chap. 1, where the wolf and deer population in a forest oscillate the amplitude and wavelength of the oscillations may be determined by the initial number of wolves and deer in the system. In addition, a random condition like the scarcity of grass and other foraging material due to a severe winter or a dry summer can change the balance of their population in a significant way.

Thus, systems are sensitive to initial conditions, where seemingly small variations can result in large variations later. Additionally, a relatively insignificant change in the environment in which a system is operating can significantly change its outcome. Rebooting or restarting is often recommended for fixing unexpected behavior of engineered systems, such as a computer or a software package. There the act of resetting the initial conditions makes the system behave in a more predictable fashion. For nonengineered systems, that type of rebooting is often impractical or impossible. However, filing for bankruptcy by a faltering company, commonly known as Chapter 11 in the USA, is somewhat similar to rebooting, which allows a business to restart with a cleaner slate. Another example of restart is a natural disaster like a forest fire, which destroys all the existing vegetation allowing a new start of natural growth.

5

Modeling and Simulation

Abstract A model represents something real or abstract. From our birth, we build mental models of things we encounter in our daily lives. Children play with toys that are models of various real objects. In our childhood, we enjoyed building models using Lego or Erector sets. Today, it is within our grasp to build interactive models on our personal computers that can simulate problems we encounter in our personal and professional lives. Interactive models help us to understand the working of systems better than is possible with words or illustrations. They allow us to build better systems and to optimize those that are in existence.

This chapter discusses the importance of modeling and simulation for understanding system behavior. It illustrates various mental, conceptual, and interactive models. It describes some of the software packages for modeling and simulation that are available in the market. It illustrates the simulation of systems, such as first-order lag, auto inventory in a dealership, and youth violence. Finally, it discusses artificial neural network for model building and the various uses of modeling and simulation.

5.1 What Is a Model?

The word "model" has many different connotations that vary between persons and situations. Children play with model trains and airplanes, which may be physically similar but much smaller and they work very differently

A. Ghosh, *Dynamic Systems for Everyone*,
DOI 10.1007/978-3-319-43943-3_5

from real objects. On the other hand, automobile and other manufacturers build models that are prototypes to try them out before starting full-scale productions. The Oxford English dictionary defines it as "A simplified (often mathematical) description of a system to assist calculations and predictions," which fits well from our system perspective.

Thus, models are descriptions or representations of something material or abstract. They are simplified representations of real systems intended to aid in their understanding. Simulation, on the other hand, is the operation of a model, enabling one to observe the behavior of the modeled system under normal and abnormal conditions and its behavior over longer time scales than is practically possible with a real system.

Engineers and social scientists often use models to represent some selected aspects of a real system or a system to be designed or built. These models are then used to simulate their behavior for better understanding, designing, implementation, and improvements of those systems. Yet there is nothing new about modeling. We carry many different models in our heads; they include our view of our immediate environment, our relationships with our friends, colleagues, and relatives, or a simple route map from our home to a grocery store. However, as mechanical, industrial, and social systems get more complex, it becomes more difficult to generate and remember those mental models accurately and more difficult to discern how they work and in what way they affect us.

With dramatic increase in computing power and its corresponding decrease in cost, we are now able to generate and use rigorous models that require little knowledge of higher mathematics. Modeling and simulation are allowing us to study existing systems, building new systems, test and optimize these systems, as well as to train people that need to work with these systems. Models offer the additional benefit of providing greater appreciation and understanding of interactions between systems and their subsystems.

Models are now used for teaching and research in many disciplines, ranging from astrophysics to weather forecasting. Economic models are used to understand and predict the behavior of past, present, and future of various economic systems. In engineering, they are used for designing, building, testing, training, and for controlling systems. An expert in any discipline can now focus on a system he or she is trying to model rather than worrying about mathematical equations.

The Advantages of Rigorous Modeling and Simulation

- Trying out a system before it is built
- Better understanding the interactions between various subsystems
- Improving the control of a system
- Improving and optimizing the performance of a system
- Learning system dynamics
- Training to operate or use a system
- Checking and reducing the possibilities of unintended consequences

5.2 Mental Models

We receive information, analyze them in our head, and respond accordingly. This type of information processing is an example of the use of mental models. Many of our mental models are based on conscious thoughts and memories, which we use constantly for our everyday activities and for longer-range decision-makings. Some of them we update almost constantly, as we get new information, such as a new or a quicker way to get to work or the grocery store that is offering better discounts this week. However, many of our mental models are unconscious and unspoken assumptions, such as our religious and political beliefs. We often treat them as our prized possessions and are not willing to modify them except under extreme circumstances.

Human beings are generally quite efficient in generating mental models. We learn to make them from our early childhood on—even as babies. Cognitive development is largely about learning to create models in our head. Thus, an infant builds an understanding of himself or herself and the reality around her along with how things work through visual, manual, and other interactions. Our educational system, both formal and informal, is geared towards the ability to create sophisticated models quickly and efficiently.

However, we store only a limited description of that external world. Only a few of us have photographic memory enabling us to remember things literally and exactly. Even if we do, it is more limited than most of us believe. Additionally, we frequently fail to generate accurate and efficient models when a system is nonlinear or when the number of system variables is more than three or four. According to Jay Forrester "Mental models of social and biological systems, which are often nonlinear, cannot be done in one's head even by a gifted mathematician." That leads us to create and stick to very simple models for complex situations, which may take us quite off the track. People in general and particularly those in power, such as politicians and legislators, often make decisions based on simple mental models, which often lead to unintended consequences or even major disasters.

Mental models, except the simplest ones, are difficult to convey from one person to another. They are context sensitive and they superimpose on our existing mental models. That is, any new set of information is overlaid on our existing models and thus is colored by our previous notions and prejudices. Yet one cannot overemphasize the extent to which we use mental models in our daily lives. We generate them quickly and they usually lead to desirable actions for our normal day-to-day activities.

5.3 Conceptual Models

The term "Conceptual model" has different connotations to different people and in various professions. Here I will define it as a type of model that is more formal than a mental model, such as a written description, an illustration, or a diagram that conveys the functioning of a system but may fall short of giving a full formal account. Thus, if I would like to get a house built, I may describe that as a list of requirements (Table 5.1). That list is not enough for a builder to build the house but may be sufficient for an architect to generate a set of formal drawings (architectural model) based on my requirements.

The conceptual model of a fruit juice blender is another example, where it may be in the form of a list of general requirements (Table 5.2), which will allow an engineer to create a detailed design of the system.

Table 5.1 A conceptual model of a house

List of requirements for a residential building
• Four large bedrooms
• Two and half toilets
• A large family room
• A study
• A formal dining room
• Kitchen and dining area
• Laundry room with a sink
• Attached two-car garage
• Approximate total area: 2500 ft^2
• Energy-efficient design with solar panels for hot water supply and large south-facing windows

Table 5.2 A conceptual model of a fruit juice blender

Functional requirements for a fruit juice blender
• Given: Two tanks for apple and grape juice
• Design and build a blender to deliver a blended juice
• Ability to blend as needed
• Flow rate up to 15 gallons/min
• Blended mixture should contain between 60 % and 62 % apple juice
• Maximum allowable error is less than 2.0 %
• Should work in environmental temperature range of 40–70 °F

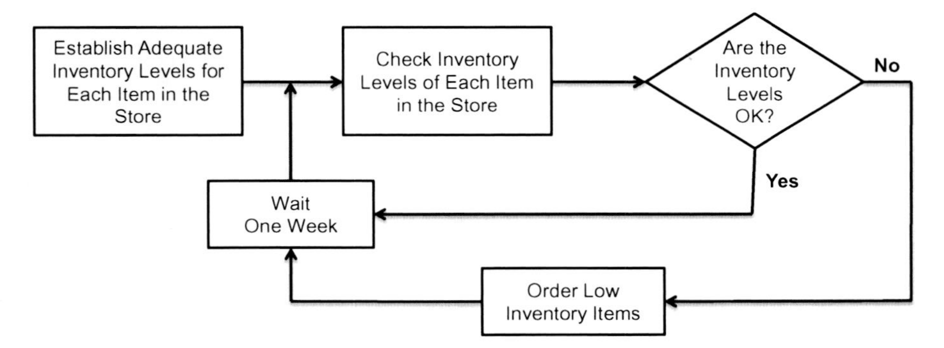

Fig. 5.1 Block diagram of an inventory control system

Conceptual models may be represented in many different ways, such as a table of requirements as shown above, or as hand-drawn sketch, graph, plain narrative, block diagram, causal loop diagram, or flowchart. The emphasis here is to give a top-level view of a set of subsystems and their interrelations. The causal loop diagrams, as shown in earlier chapters, may be considered as conceptual models. Alternatively, a block diagram may be appropriate for a conceptual model for maintaining the inventory level for a store (Fig. 5.1).

Conceptual models are used not only for building systems but also for studying systems that are in existence, such as a population growth table for different countries in this world or emission of greenhouse gases in a region. Conceptual models may also be called by many different names, such as requirement specification or functional model. A conceptual model is also useful for testing the functionality of a system to be built and is often a stepping-stone for generating models that are more rigorous.

5.4 An Interactive Model Is Worth a Thousand Pictures

With an interactive model, one is able to change input and other variables to observe the resultant effects. Thus, it is able to simulate various normal and abnormal situations based on users' inputs. That makes an interactive model a far more effective vehicle for learning than words, pictures, or live videos. There is a common saying "A picture is worth a thousand words." The idea being that a complex thought can be conveyed with just a single still image. To that we may add, *An interactive model is worth a thousand pictures.* That aptly describes one of the main roles of an interactive model, which allows us to understand and experience real-life situations much more efficiently than is possible with pictures and illustrations.

In principle, an interactive model may be a hydraulic, mechanical, or an electromechanical device, such as those used for training of astronauts or airline pilots. Now, with the increased availability of simulation packages, software based modeling is gaining ground.

A rigorous model may be built using detailed knowledge of system components and their interactions, which for mechanical systems may include physical variables, such as energy, mass, power, and storage capacity and for biological systems, health, environment, and reproductive power. For an institution like a school, that may include facilities such as classroom and playing fields, quality of teachers, motivation of students, and the like.

Many software packages are now available for simulation of systems both large and small. Some are for very specific applications, such as weather forecasting, optimization of a manufacturing process, or for training of airline pilots. Many of them run on large mainframe computers; however, there are also a number of software packages that are suitable for modeling and simulation of common systems that run on personal computers. They include STELLA/iThink, Insight Maker, MATLAB/Simulink, and many others. More details on these packages are included in Appendix IV.

STELLA and iThink packages are almost identical. They depict a system by stocks, flows, converters, and connectors (Fig. 5.2), where:

- A stock is a reservoir that accumulates, which is represented by a box or a rectangle. Examples of stocks may be the number of people in a city, amount of grain in a silo, or gross national product of a country. The value of a stock is dynamic that is, it may vary with time depending on its inflows and outflows

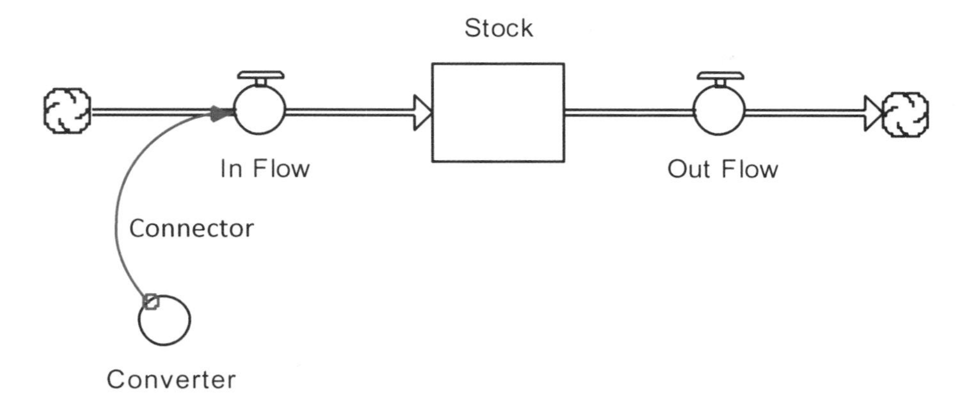

Fig. 5.2 Basic building blocks of STELLA

- A flow is depicted by an arrow-headed pipe that goes in or out of a stock. An arrow has a tap or a faucet that can be turned to increase, decrease, or completely turn off the flow. A cloud at the end of a flow indicates that the source or the sink of the flow is outside the system under consideration
- A converter is represented by a circle that may serve multiple functions, such as storing constants, variables, or equations. When a converter has only inputs or outputs then it acts as a repository of information or data. When it has both inputs and outputs, it converts the inputs to generate the output using stored constants or equations. The populations of a country, interest rate of a bank, desired temperature of a building, or rate of change of speed of a motor vehicle are examples of information that are stored in a converter
- Connectors transmit information between stocks, flows, and converters

Insight Maker is a web-based modeling and simulation package, whose user interfaces are similar to that of STELLA. Its website includes a large number of prebuilt examples of models. A user does not have to pay any fee to use this package, making it ideal for a beginner.

Simulink is a set of software tools for modeling, simulating, and analyzing dynamic systems. Simulink offers a graphical block diagramming tool and a library of function blocks that may be linked together to simulate a system (Fig. 5.3). It is a subset of MATLAB package offered by MathWorks and it offers close integration with the rest of MATLAB environment.

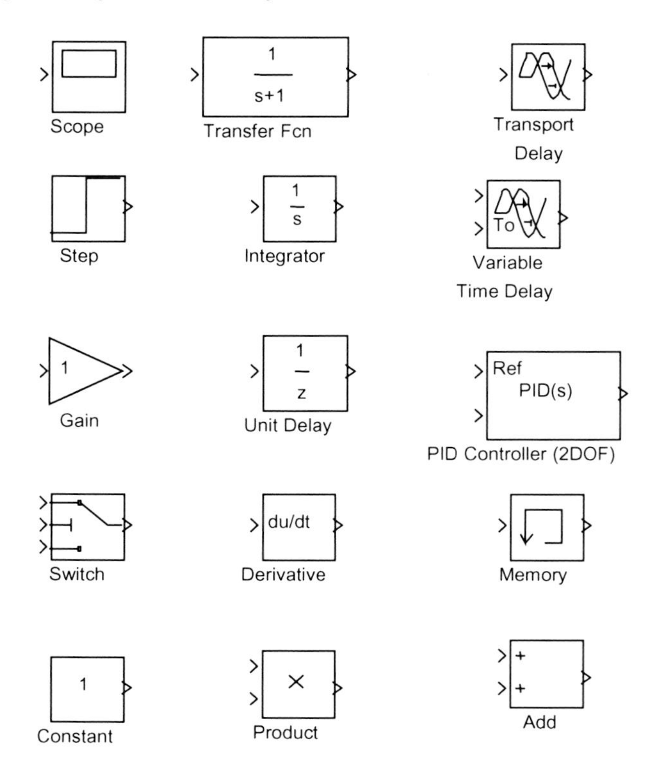

Fig. 5.3 A representative list of Simulink blocks

5.4.1 Simulation of a First-Order Lag

In Chap. 2, we discussed first-order lag using the example of an open water tank. Here, water outlet flow depends on tank level and the amount of opening of outlet valve. If the outlet valve is kept in a fixed position, then the outlet flow will depend only on the height of water in the tank, where an increase in water level in the tank will lead to a greater outlet flow. That may be simulated by using various software packages (Fig. 5.4).

Suppose the system is in equilibrium, where the water in the tank is at a certain level and the input and output flows are equal. Then, a sudden increase in water inflow rate will cause the tank level to rise resulting in an increase of outlet flow until the system comes to another equilibrium state, with a higher tank water level (Fig. 5.5). Here the outflow of water shows a first-order lag.

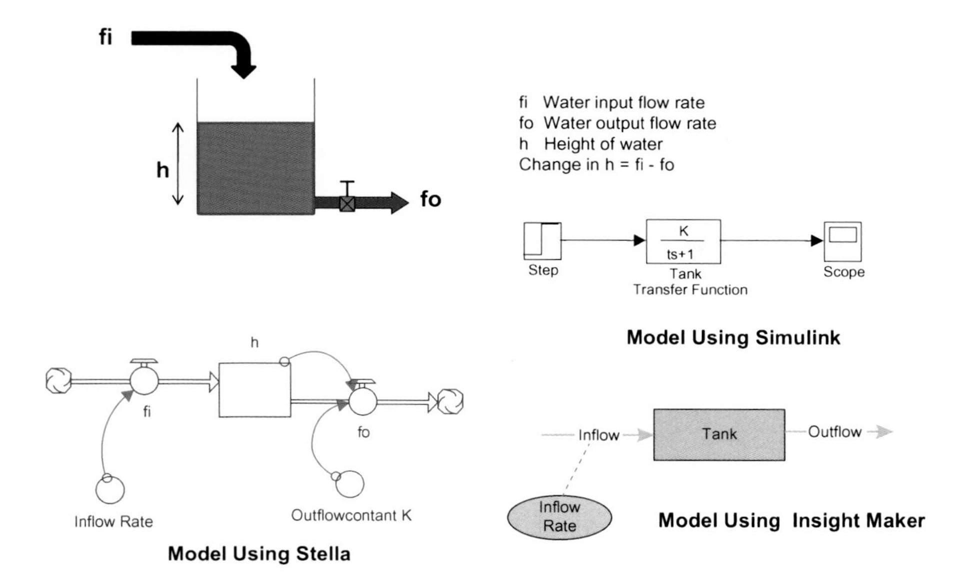

Fig. 5.4 Modeling a first-order lag

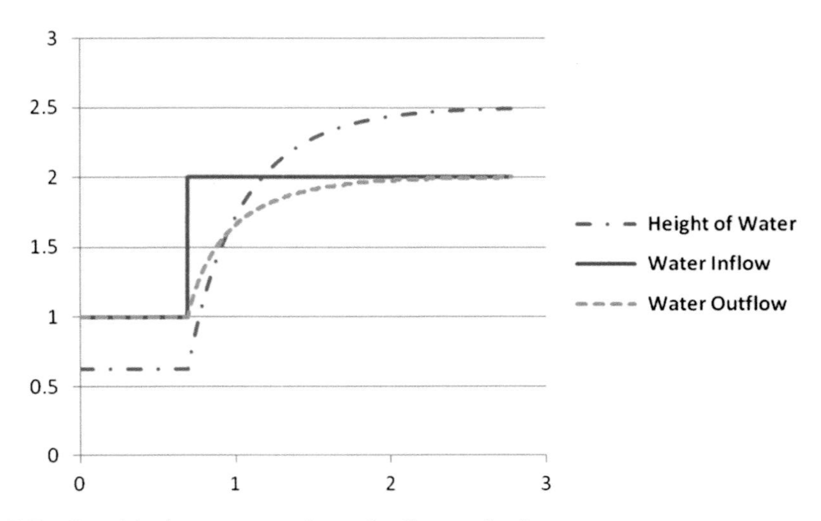

Fig. 5.5 Graphical representation of a first-order lag

5.4.2 Simulation of an Auto Dealership's Inventory and Sales

In the last chapter, we briefly discussed the case of an auto dealership where delay in new auto delivery from a manufacturer results in oscillation of the dealership's inventory level. For simulation, let us suppose that the optimum inventory level in the dealership is 1000 cars and the sales volume in any given month is directly proportional to the inventory. Let us assume the sales volume on any given day is 20 % of the inventory level. The dealer places order to the automobile manufacturer for new cars as soon as the inventory goes below the threshold of 1000, where the order volume is equal to the amount of shortfall. This may be simulated by STELLA (Fig. 5.6), where the stock is Auto Inventory and the inflow is New Auto Delivery, which is based on the shortfall and a variable Supply Lag. The outflow is the Sales Rate, which is 20 % of the inventory and the time scale is in months. The simulation results show that there is little oscillation when the lag is around 2 weeks (half a month) and a damped oscillation when it is around 1 month (Fig. 5.7). However, there is a sustained oscillation of the inventory level when the delivery time increases to one and a half months.

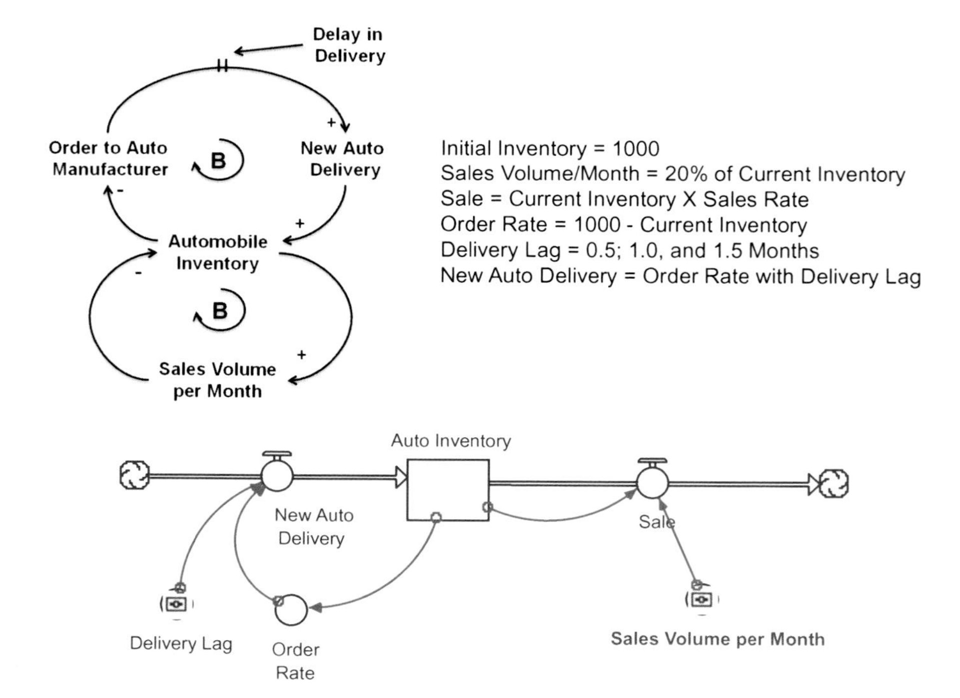

Fig. 5.6 Modeling and simulation of an auto inventory

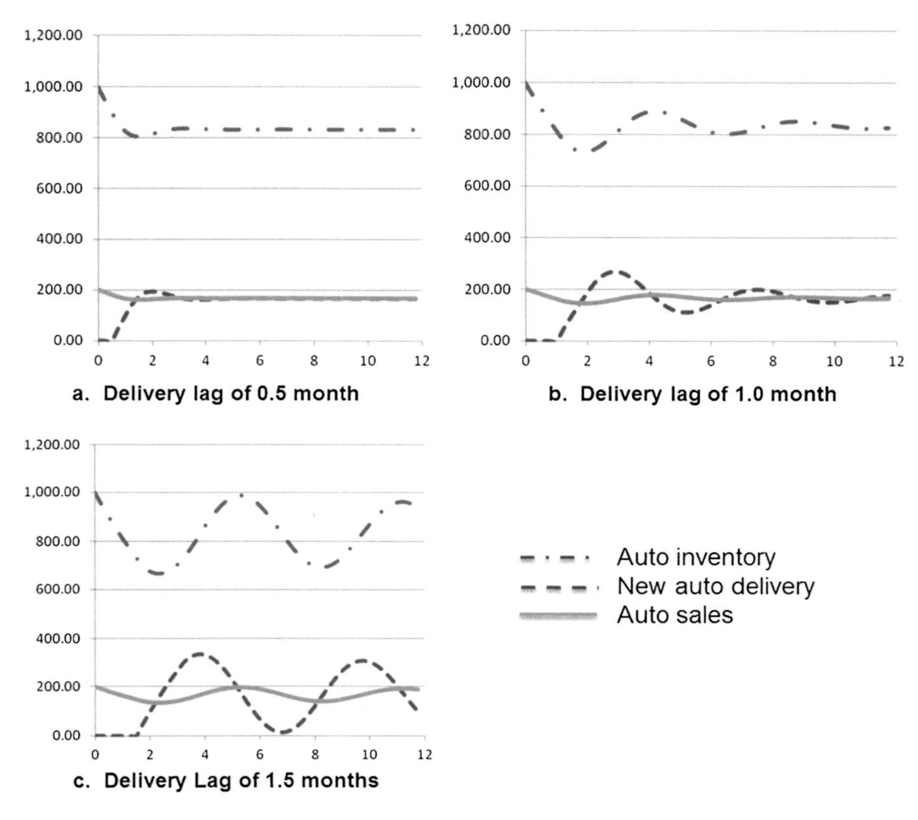

Fig. 5.7 Auto inventory simulation results

Here, oscillation in the inventory level can be reduced by improving delivery time of new automobiles, but that is under the control of the manufacturers on which the automobile dealer may have little influence. The dealer may preorder vehicles a month in advance based on anticipate sale volume, thus reducing delivery lag and the resultant oscillation. That will of course increase inventory cost, which is a trade-off for the dealer. Thus, the simulation clearly demonstrates the negative effects of delays in a system.

5.5 An Interactive Model for Youth Violence

In Chap. 3, we discussed the dynamics of youth violence in a Boston neighborhood and generated a number of causal loop diagrams to illustrate it. Here, we will discuss the benefits of creating an interactive model based on the comprehensive loop diagram generated earlier.

5.5.1 Simulation of Youth Violence

The Youth Violence System Project (YVSP) team used STELLA/iThink software package to generate the simulation model (Fig. 5.8). In developing the model, the builders endeavored to develop an objective and value-neutral representation of the relationships that stimulated youth violence. The model used stocks and flows to account for changes in the composition of a community's youth population over time.

The interactive model allowed users to simulate many possible scenarios by varying the rate of flow of persons moving between various groups and to observe the results of these variations. Additionally, the model allowed users to vary common factors, such as the rate of youth movements from "Uninvolved" to "Associate" and then to "On the Edge" or the rate of imprisonment of those that are involved in violent activities.

A standard facility of the simulation package is the ease of generating graphical outputs. For example, it was quite easy to simulate and compare the effects of late and early interventions (Figs. 5.9 and 5.10). Here, early intervention included effective measures to reduce the Associates from moving to the Edge. That could be realized by providing better sports and recreational facilities along with jobs during vacations and after graduation. Late interventions took the form of higher rates of imprisonments and near-zero tolerance to unsocial or criminal behaviors. The, simulation results indicated that early

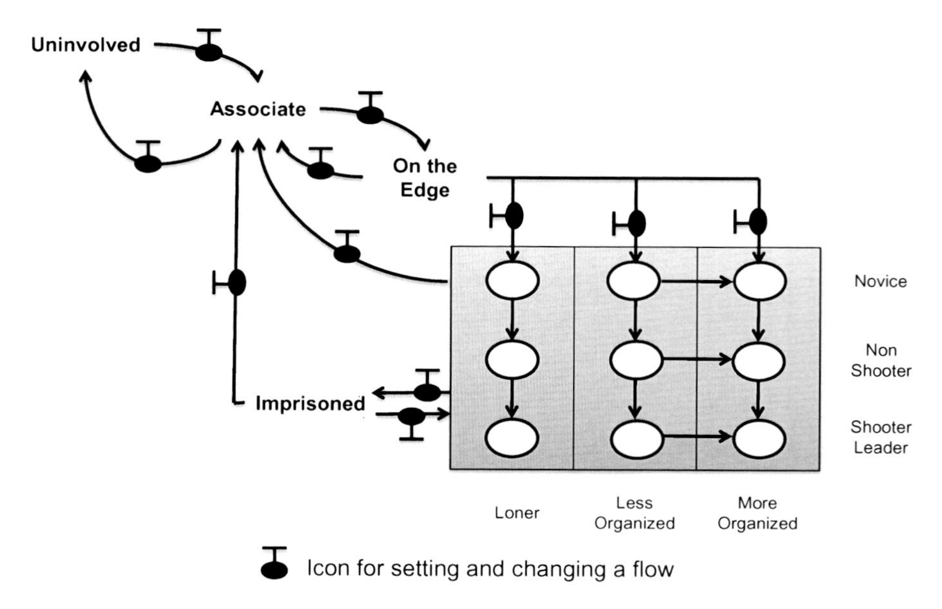

Fig. 5.8 The simulation of the paths to violence

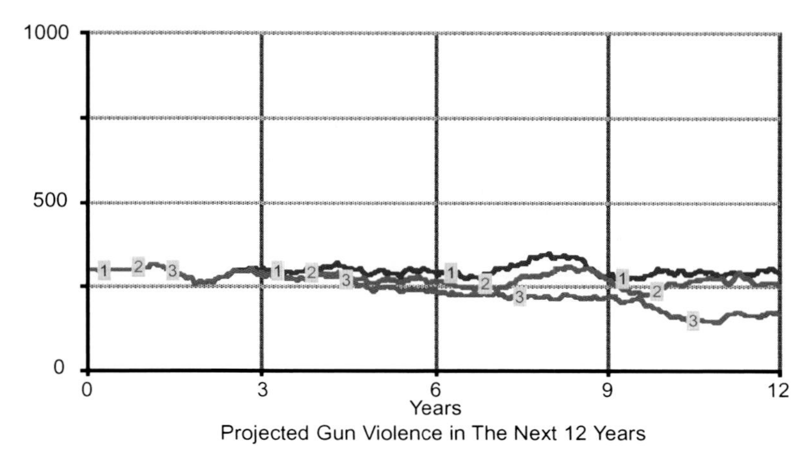

Fig. 5.9 Effects of late and early interventions on gun violence

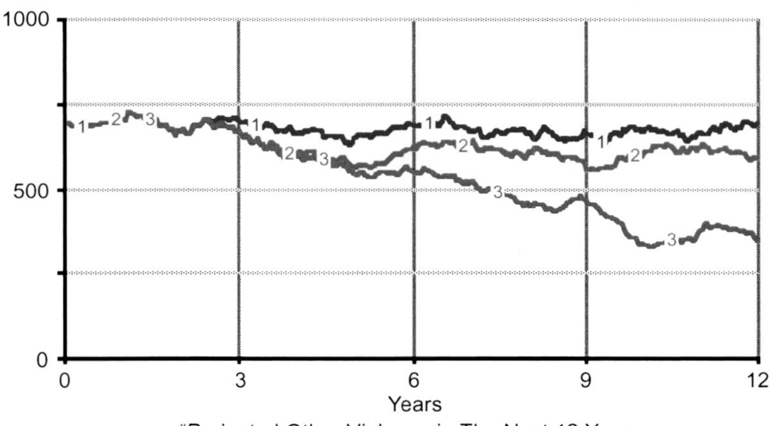

Fig. 5.10 Effects of late and early interventions on other violent activities

interventions are more effective than late interventions in reducing gun and other forms of violence.

5.5.2 Project Benefits

YVSP employed a community-based participatory process in which residents participated in design, execution, and evaluation of a detailed, computer model for youth violence. Community leaders, uninvolved youths, and gang members provided unique insights into the behaviors of violent youths in their neighborhoods. Empowerment and engagement of these groups fostered a collaborative environment in which the logic of community residents were articulated and explored.

The use of a group approach to generate the system dynamics model created enthusiasm from community residents because they saw their own logic reflected in the evaluation of violence-reducing initiatives. This community-based approach coupled with system dynamics methodology produced a new understanding of youth violence. Most of the participants experienced significant changes in perception towards the causes for youth violence, which include:

Traditional Perceptions

- Concentrated poverty and inequality
- Street codes and the culture of violence
- Fatherlessness
- Family decline
- Community tolerance
- Young people join gangs for: protection, respect, fun, money, and acceptance

New and Modified Perceptions

- Escalation of violence is caused by increase in gang members
  - The relationship is nonlinear
  - In a high-violence neighborhood almost every youth has some connection with gang members
- Traumatic stress changes the way nonviolent youths behave
  - Increased stress when violence happens in close proximities
- Gang violence is addictive
  - Violence is a way to release tensions
  - Killing is like taking a drug, there is a feeling of "rush" associated with violence
- Significant unintended consequences of suppression strategies
  - Longer-term sentences are more effective than multiple shorter-term sentences
- Gangs represent a significant alternative social system
  - Gang affiliations often supersedes familial and cultural associations
  - Often, there is no remorse in killing biological relatives
  - Often it is a tribe without elders

Thus, the system dynamics methodology and community-based approach produced a better understanding of the problems associated with youth violence. The model gave a more accurate description of the behavior of youths in high-violence neighborhoods than was possible before. Interactive simulation of various if–then scenarios allowed the identification of effective leverage points to optimize the benefits of interventions. However, the generation of an interactive model for a problem like youth violence and applying its results in a society may take a long time to produce tangible benefits.

5.6 Black Box or Empirical Models

So far, we have discussed the modeling of systems whose inner workings are known or well understood. These models are generally called first principle or white box models.

However, there are many other situations where detailed knowledge of the inner workings of a system is not available or are too complex. A model for such a system may be created by observing the system's external behaviors that is by correlating its outputs with its corresponding inputs. Such a model is termed as black box or empirical model. Most of the mental models that we generate tend to be empirical, like understanding a person by observing how that person behaves under various stress situations or trying out various medications to get relief from a chronic medical condition. Artificial neural network (ANN) is a preferred way of creating a model of a system whose inner workings are not well understood.

5.6.1 Artificial Neural Network

ANN emulates biological neural system in our brain (Fig. 5.11). The key element of ANN is an artificial neuron, which may have multiple inputs and outputs that are connected to other such elements. However, an input neuron, that is, one that gets signal from the outside world has only one input, while an output neuron has only one output. Each connection, depicted by an arrow, has a weighing function, which acts as a multiplier. That is, higher the weighing function stronger is the signal.

A typical network may consist of a set of input and output neurons and a number of hidden ones that are positioned in between. The hidden neurons may be in single or multiple layers.

ANN, like our brain, learns by examples. During learning mode a set of variables are fed to the network, which in turn produces a set of outputs.

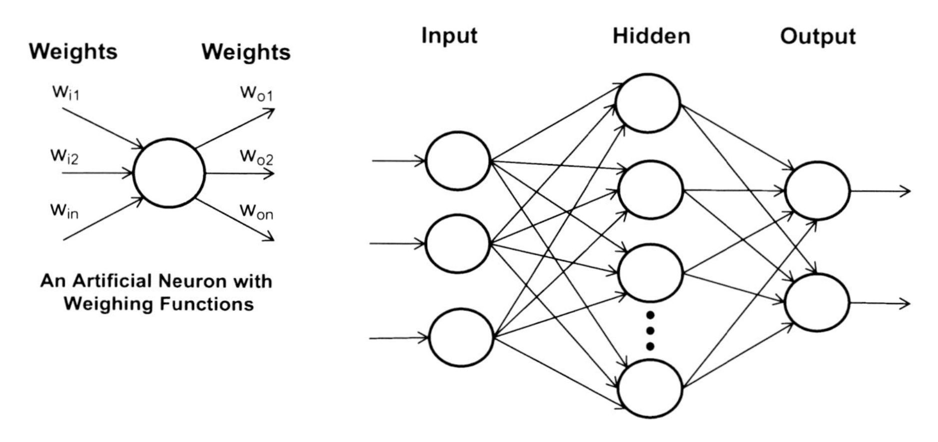

Fig. 5.11 Artificial neural network

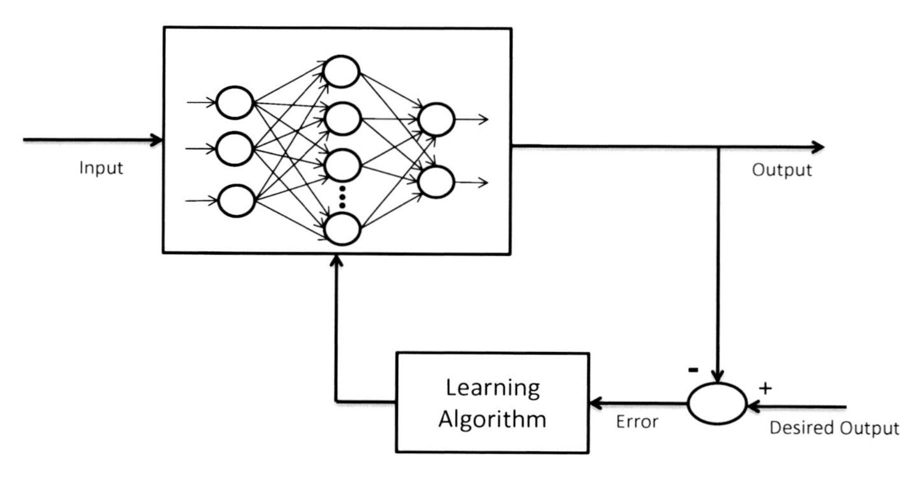

Fig. 5.12 An artificial neural network in a learning mode

A comparator compares the actual outputs with a set of desired outputs to calculate the errors. These errors are then used by the learning algorithm to adjust input and output weights of neurons to reduce the errors. The process is repeated until the error is minimized. The whole process is then repeated with another set of input variables and so on until the network has learned enough within the domain it is supposed to operate. This ability to learn from examples allows ANN to generate a black box model of a system (Fig. 5.12).

Neural networks and conventional modeling methods complement each other. As stated before, conventional or white box modeling is well suited for systems whose internal functions are reasonably well understood, whereas neural network may be used when they are not. Additionally, there are a large number of situations where a combination of these two approaches is more appropriate.

As stated before, an ANN needs to be trained for a number of different scenarios within the area it is supposed to work. Training for each of these situations may take a significant amount of time, making the total training time quite long.

Advantages of ANN

- Not necessary to know the inner workings of a system
- A network can be used to analyze new conditions
- Can learn from noisy and incomplete data
- Suitable to solve complex problems

Disadvantages of ANN

- The model is not interpretable; it is a black box
- Cannot completely replace conventional modeling methods
- Require lengthy training and testing times
- Require a huge number of artificial neurons to simulate a real system (a human brain has around 86 billion neurons)
- Accuracy of the results varies with its architecture
- At present, NN computers are quite expensive.

Neural Networks Application Examples

- Pattern recognition (i.e., radar systems, face identification, object recognition)
- Game playing and decision-making (i.e., backgammon, chess, racing)
- Sequence recognition (i.e., gesture, speech, handwritten text)
- Industrial process control
- Robotics
- Medical diagnosis
- Financial applications (i.e., stock market prediction, automated trading)

Artificial neural networks had initial successes, in 1960s, when they were found to be useful in some limited areas of application in pattern recognition and industrial process control. However, it lost its appeal for wider applications as the computing power available then allowed emulation of only a limited number of neurons. A typical network in those days had a few hundred neurons and they were usually organized in single or a couple of layers. Today, one is able to simulate billions of neurons organized into a number of hierarchical layers, where each layer deals with a different level of abstraction. That makes it more useful for many real-life applications, such as image recognition, financial forecasting, and robotics. Such a system, however, is expensive for common use and needs a long period training. Wider use of ANN will come about only when they become more cost effective.

5.7 Models for Simulation and Training

Use of flight simulators for training of airline pilots and astronauts is well known, but there are now many other applications of models for education and training. They range from simulators for the training of police force, army, operators of large industrial systems, such as oil refineries, to teaching of mathematics, and other physical sciences in our schools and colleges. They mimic aspects of real-life situations, either completely or partially, as needed for an application. Some of these simulators are purely software-based, that is they use computer programs along with computer-generated displays for manual interactions. Others such as flight simulators use electromechanical devices along with computer models to simulate real-life situations.

However, a model is not a real system, so its actions are limited by the scope of its design. Model building is both art and science. Thus, a good model should be able to simulate those functions of a real system that are of primary importance for a given application. For example, the model for designing and building a chemical plant is not identical to the model for training a person to operate that plant.

5.7.1 Training Astronauts

Modeling and simulation are extensively used in the training of astronauts. That includes fixed-base simulators, which do not move but operate in every other way exactly like Space Shuttles to give the crews the feel of on-orbit operations. A motion-based trainer, on the other hand, simulates vibrations, noise, and views the astronauts experience during shuttle launch and landing.

In a virtual reality laboratory, astronauts prepare for spacewalks or robotic arm operations. In a simulated environment generated by computers, astronauts learn how to orient themselves in outer space, where zero gravity allows them to move in three dimensions. Prospective astronauts spend long hours on these simulators to prepare themselves for a variety of contingency situations. Such is the power of simulation that astronauts often say that real space flights are often easier than the long and tough time they had with the simulators.

5.7.2 Training Industrial Plant Operators

Process modeling and simulation are increasingly being used in the training of operators in large process manufacturing industries, such as for oil refineries and chemical plants. Operator training simulators are allowing new and inexperienced personnel to gain knowledge and experience to operate them quickly and effectively. Often these simulators are built and used before the plants go into production, allowing faster, smoother, and problem-free start-ups.

A strong model-based training program helps eliminate errors that can lead to accidents, product loss, or costly plant shutdowns. Continuing refresher training helps keep the skills of operators up to date for infrequently performed tasks such as start-ups and shutdowns, and in dealing with unexpected events. There are significant savings as better training reduces start-up time. Model-based training is also offering significant benefits to industries such as pharmaceuticals, food and beverage, and life sciences, which must conform to tight training regulations.

Model-based training is becoming more and more an integral part of an operating facility, not just to train, but also to do "what if" scenarios, troubleshooting, engineering studies, and to enhance continuous improvement efforts. It is, however, imperative that these models are kept up-to-date with the changes made in plant equipment or control configurations.

The concept of virtual reality is now gaining ground where a model mimics a real-life situation allowing a person to interact and experience an environment in a realistic fashion using various sensory and motor organs such as eye, ear, skin, hand, and foot. Viewed against some of the latest advances in interactive gaming technology, today's operator training systems are still in their infancy. The training of plant operators in future will include simulation of various normal and emergency scenarios based on virtual reality and testing trainees on their ability to cope effectively both procedurally and in timely fashion.

5.7.3 Training in an Academic Environment

While learning mathematics in a high school, students usually find little time to develop the underlying concepts. The abstract nature of mathematical representations can easily amplify students' difficulties. Converting word problems into appropriate equation format often entails a conceptual leap that challenges even the better students. There, a modeling software tool can dramatically increase conceptual learning process, where its graphical interface makes it easier to visualize mathematical relationships and allows students to observe the solutions for a variety of conditions.

A model is a bridge between the words that describe a concept or a problem and its formal mathematical representation. Thus, a deeper level of understanding of a mathematical problem is achieved by using models as a part of the teaching process. Teachers in a mathematics department often report that modeling greatly improves their teaching productivity. As an added benefit, the teachers find that the students, with their knowledge of modeling and simulation, are often better prepared to assimilate concepts presented in other courses in the natural sciences. Similar benefits are accrued in other fields of study, such as business, economics, psychology, medicine, and public health.

5.8 Importance of Interactive Modeling and Simulation

Building a model is of little use unless it provides some insight beyond what is already known from direct investigation of a system under study. Thus, one of the main justifications for modeling is to increase the fundamental understanding of a system. While mental models are good and efficient for simple situations with small number of independent variables, they fail to predict outcome when there are a number of interrelated inputs and variables. As most of the real world systems are nonlinear, the common assumptions of linearity in mental models often produce erroneous results. There, an interactive or live model with the ability to simulate a large number of variables and their nonlinear behaviors allows us to generate various possible scenarios that could not be foreseen otherwise.

Rigorous modeling and simulation has been the domain of technologists and experts but that is now changing. With increased availability of PC based modeling and simulation software, its use will increase dramatically in coming years. Following narrative, summarize the main advantages and limitations of modeling and simulation.

Limitations of a Model A model is not an actual thing but an abstraction of certain parts of a real system. A model, mental or otherwise, is only as good as the data and the relational values that are used to build them. When that is incomplete or inaccurate, the model does not give a true picture.

Since it is only an approximation of a real system, it is inherently inexact. Additionally, initial or starting conditions of a system, which may affect its behavior significantly, may vary producing results that are different from an actual system. Moreover, all models have operating limits. Beyond those, the behavior of a system may be erroneous or impossible to predict.

Despite all these weaknesses, models are very powerful tools for the understanding of many types of systems. Often a model is the only viable way to extrapolate the future of a large or a complex system. Thus, when using a model, one needs to understand and appreciate both its capabilities and its limitations.

Ability to Try Out a System Before It Is Built Most systems are expensive in manpower and material to build. As the cost of failure of a large system is quite expensive, it was a common practice to build it in a smaller-scale (pilot plant) and then scale it up once it is well proven. With the advent of computer modeling and simulation there is lesser need for the construction of such pilot plants.

Improving the Performance of an Existing System A system may be working reasonably well or it may have some specific problems. In either case, one can alter some of its settings to observe the results. However, that may be time consuming and hazardous. It is much better if that is tried out on a model first before applying it to a real system.

A Learning Tool for System Science System science deals with interactions, which are not always easy to describe using English narratives or even with pictures or graphs. These interactive models act as powerful teaching tools. By building and using models one gets better insight into interdependencies in our lives and in systems around us.

Designing a System with Little Mathematical Knowledge Proper design of a large or complex system requires the knowledge of all its subsystems and their interactions. It may be possible to develop mathematical equations describing those subsystems and their interactions, but that will require considerable mathematical skills and even then, it may be difficult to predict all possible situations that the system may encounter. Building a model based on a readily available software package, a builder can focus on its functional aspects without spending a lot of time and effort in mathematical formulations.

Reducing the Possibilities of Unintended Consequences With an interactive model many possible scenarios can be simulated thus allowing the testing of unusual and extreme conditions. That makes it easier to detect and analyze situations that are unintended or hazardous.

Training to Operate or Use a System

In addition to training astronauts and airline pilots, models can be used in many other fields, such as police training, military combat training, advanced drivers' training of motor vehicles, operator training for oil refineries and petrochemical plants. The use of simulators for training purposes is exploding.

Modeling as an Engineering Tool Process modeling and simulation has demonstrated its value as a tool for designing plants and processes. As in an engineering application, simulation plays a significant role in shortening engineering projects and improving the quality of design. It is also playing an increasingly important role in manufacturing operations as a decision-supporting tool.

6

Optimization

Abstract Optimization is the process of finding the best possible solution for a given problem. This is simple when there is only one variable that needs to be optimized. For example, if one would like to find the shortest route to work, then one needs to focus only on driving distance. However, the problem gets more complicated when there is more than one variable that needs to be considered, such as the shortest route along with minimum driving time and fuel cost. There one has to make some compromise between these multiple objectives to come up with an optimal solution.

This chapter outlines the challenges of optimizing a system with multiple goals and various constraints. It discusses some of the methods, such as model-based control, hill climbing techniques, and kaizen (optimization in small steps). The chapter includes a number of examples, such as optimizing work force requirements for a project, reducing energy consumption in a home, and an objective decision-making process. Finally, there is a discussion on the various opportunities and challenges for optimization.

6.1 What Is Optimization?

Optimization comes from the word optimum, which is defined as "the best or the most favorable condition or situation." Thus, to optimize a system means making it to operate in a way that produces the most desirable results. In economics and social sciences, "to optimize," means choosing the best option from a set of possible alternatives.

A. Ghosh, *Dynamic Systems for Everyone*,
DOI 10.1007/978-3-319-43943-3_6

For engineered systems, optimization means maintaining a set of conditions to achieve the most advantageous operation of a system. For example, there are a number of ways to operate an electric power plant, where one has to decide what constitutes its optimal performance. Those include maximum thermal efficiency, lowest possible emission, lowest possible cost, and maximum system availability. These goals are interrelated but in some cases, they are at odds with each other. For example, high excess air will result in better carbon burnout leading to less smoke and carbon monoxide production, but that will also result in lower thermal efficiency and higher emissions of harmful nitrogen oxides. These interactions need to be taken into account when trying to optimize such a system.

The task of optimization is simple when there is only one variable that needs to be controlled. This is a setpoint control problem, as was the case for controlling the thickness of steel plates in a rolling mill in Chap. 1. There, the thickness of rolled steel plate is the only variable that needs to be controlled even though there are a number of factors, such as temperature of the heated steel slab, clearance between each pair of rollers, and the speed differences between pairs of rollers that affect the outcome.

However, optimization of two or more interacting variables in a system can be quite complicated. For example, I may minimize travelling distance for my daily commute to work by choosing the shortest possible route. However, I may also like to minimize driving time. There, the shortest route may not be the quickest. I may be better off taking a highway where there are less slowdowns and stops, even if the distance is longer. The problem gets even more complex if I consider traffic situations in these different routes and their variations with time and day.

Another example of optimization would be, minimizing my monthly grocery expenses. If there are three grocery stores A, B, and C in my town, and I know that store C offers lowest prices for most items, then I can reduce my grocery bill by always shopping at store C. However, that may not be the optimum solution if store A offers certain items at a lower price than store C. Then, to optimize my grocery bills I may buy the cheaper items from store A and all other items from store C. However, the two-store option may entail extra time and travel because they are located at opposite ends of the town. In that case, I need to take into account the extra time, and increased fuel expenses and then decide whether I should buy from both stores or just stick to store C. Thus, determining the strategy for minimizing grocery expenses can be quite complex.

6.2 Optimizing Manpower for a Project

Sam, a project manager, had a big problem. The software project he was leading was behind schedule. According to his best estimate, it would take more than a year to complete the project with the present workforce. This was unacceptable to the client, who wanted it to be ready and running in 9 months or less.

Though Sam has 10 experienced engineers on his team, working full time, for several reasons, he could not schedule enough overtime to shorten completion time to 9 months. Therefore, he needed more engineers, but it was hard to find them with right background and experience.

Even if he could get right personnel, it would require considerable time and effort to train them, which would fall on the shoulders of existing team members, thus, reducing their productivity for a while. Increased team size would also significantly add to the amount of time the team would spend on communication and coordination (Fig. 6.1). Additionally, a significant increase in staffing would raise project execution cost to an unacceptable level. Sam was not certain whether adding extra staff would be a good idea, but he had not many other options. He then decided to go to the human resources (HR) department to get advice and help.

The HR manager told Sam that there were a number experienced engineers in another project that was nearing completion. He could hire some of them, but she needed to know, as soon as possible, the number of persons Sam would need.

To determine the number of engineers to add, Sam faced two competing objectives—reducing project completion time and controlling staffing expenses. He was well aware of the fact that the new engineers would require training, which would largely be provided by the existing staff members.

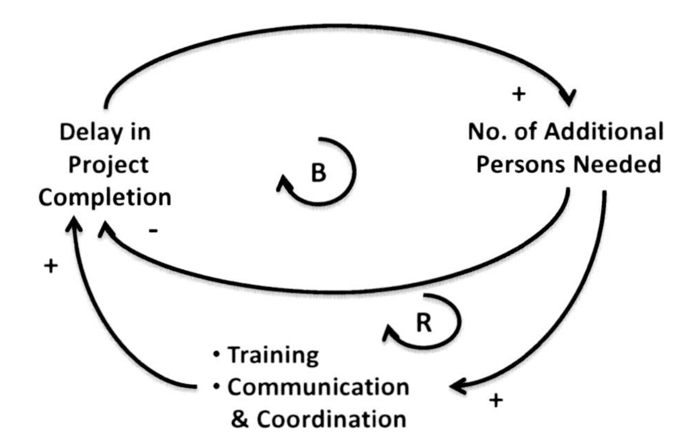

Fig. 6.1 Causal loop diagram of a project running late

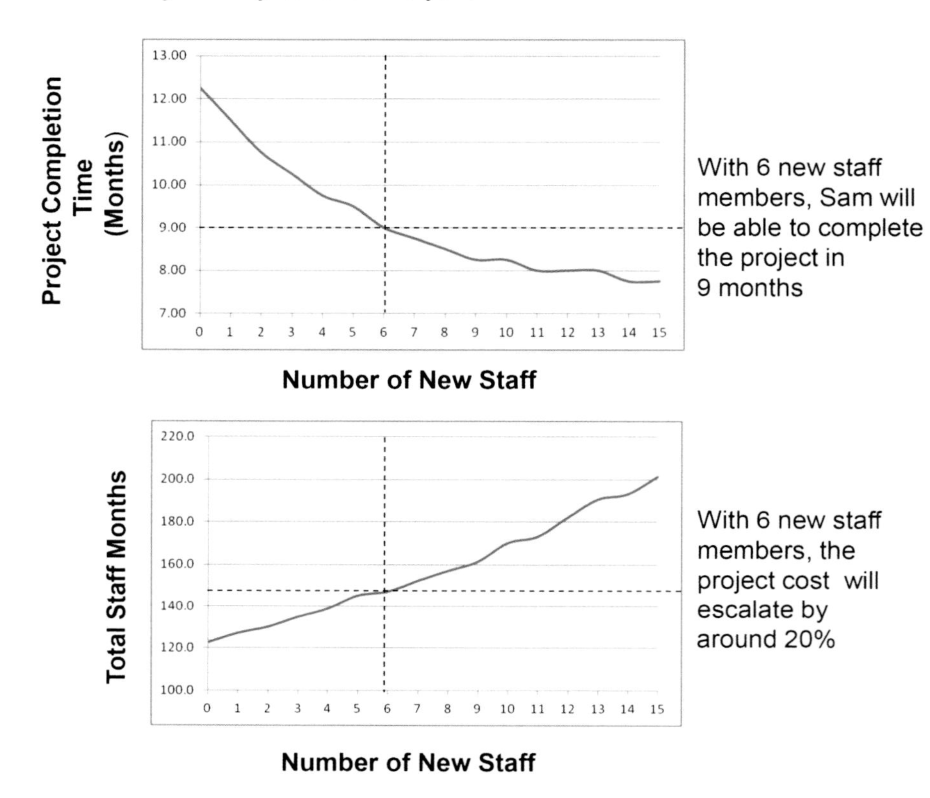

Fig. 6.2 Project completion time and total staff months

Additionally, with a larger team, there would be a significant increase in their coordination efforts. The existing team had a flat structure where team members spent considerable time in communicating with each other to ensure compatibility between the software modules they produced. With a larger team that was going to increase significantly.

He went back to his office and, after some thought, decided to build a model, which would simulate project completion time with varying number of additional staff (see Appendix V for simulation details). The simulation allowed Sam to generate two graphs that showed the effects of adding new staff on project completion time and on total staff months required for completion of the project (Fig. 6.2).

Based on the simulation, Sam made the following conclusions:

- With current staff, the project will take just over 12 months to complete.
- By adding new staff, he will be able to reduce the project completion time, but that will also increase the total man months of effort, thus, increasing the project cost.

- By adding six new members, he will be able to reduce the project completion time to 9 months.
- Then the project cost will escalate by around 20 %, which would be acceptable.

Thus, by simulating work force and completion time of the project, Sam was able to make an informed decision on optimizing the additional manpower requirements. Additionally, the model was quite general, so it would be useful for optimizing staffing requirements in many other situations.

One of the Most Important Findings
There is a significant escalation of communication and coordination efforts with the increase in personnel in a team. Thus, reducing this overhead is of paramount importance in controlling the cost of a large project. Therefore, Sam decides that in future, he will divide a large project into a number of major tasks, where each task will be tackled by a small group of engineers. Then individual engineers will interact only with members in his or her group while the group leaders will interact with the leaders of other groups. That will minimize the need for interactions between individuals in the different groups leading to a significant reduction in the communication and coordination overheads.

6.3 Objectives and Constraints

In this example of optimizing manpower requirements, the objective is to minimize manpower requirements while completing the project on or before a deadline. Here the project completion time is the constraint. However, there may be other constraints that a manager may face for a similar project, such as lack of suitable personnel or longer training time for those that are available.

Other situations may have their own sets of objectives and constraints. For example, if I am intending to buy a notebook computer, I would like to choose one that will serve my needs in the most cost-effective way. That includes its purchase price, its durability, and its maintenance cost. Thus, my objective is to minimize total cost of ownership. However, there are a number of constraints, such as maximum amount of money I can afford, minimum acceptable speed of the processor, and minimum screen size. Additionally, I do not want one that is manufactured by company X, because I had a bad experience with their products, which is another constraint. By taking into account all my constraints, it may not be too difficult to find a number of notebook computers that fit my bill. Then I can choose one that is most economical in that group. However, if there is no computer that can satisfy all my constraints, then the decision-making process will get rather complicated. There I may have to decide on the constraints that are absolute for me and those that may be modified.

These are simple examples of optimum decision-making. They are simple, because they have a single objective and a clear set of constraints. It is, however, not so simple in real life, where there may be more objectives than one. For example, in choosing the computer, I might also want high reliability along with a high quality keyboard. Then, I have multiple objectives to optimize, but I may find that there is no model in the market that is superior in all the above criteria. That spells trouble, as any solution to a problem like this will require tradeoffs between these multiple objectives. It might be advantageous to denominate all constraints in terms of dollars, e.g., I will pay US$ 100 more for a branded computer than for an off-brand one.

One way to simplify such a problem would be to consider the primary objective as the only objective and treat others as constraints. Thus, if I can quantify the minimum acceptable standards for reliability and keyboard quality, then I can eliminate those that do not meet these requirements. That will leave me with only those that are acceptable, from which I can select one that is most cost-effective in the long run.

6.3.1 Constraint Analysis

Constraint analysis or finding the barriers in a system is an important part of an optimization process. All systems have constraints but not all constraints are bad. For example, I have a limited income, so I have to put constraints on how much I can spend monthly on nonessential items. Parents may impose constraints on their children on amount of time they can spend per day in watching television. These are examples of manmade constraints. However, we face many other natural constraints in our daily lives. Scarcity of freshwater may put restrictions on using water for nonessential purposes like watering lawns or washing cars, and inclement weather can limit an airline company's ability to fly aircraft on schedule.

Not all constraints are alike. There are hard constraints, soft constraints, and many in between. Hard constraints are absolute and unalterable, such as laws of nature. Effects of gravity cannot be altered, while we are on earth's surface. Constraints related to safety may also be considered hard, for example, maximum pressure a reactor can withstand or how fast a train can travel given the condition of the track. As these absolute constraints cannot be altered, we need to take them into account while designing or optimizing a system.

However, there are many other constraints that are manmade, which may also be quite difficult to alter. The constitution of a state and some social and religious norms are examples of such constraints. Then there are soft constraints, such as social custom, behavior, and dress code for different occasions like in an office or at a beach party. Between these extremes, there are hosts of other constraints that may be surmounted by insignificant or considerable efforts and expenses.

When we are considering engineered systems, such as mechanical or electromechanical devices, their constraints can often be modified by its designers or builders, but that may not be easy for their users. If 10 s is the minimum time an automobile takes to accelerate from stationary to 60 miles per hour, then that is a hard constraint for its user. No amount of extra push to the accelerator can overcome this constraint. However, the designer of the vehicle may find it comparatively easier to alter the fuel injection system or other components of the engine to improve acceleration for future models.

In a system, a constraint may be either internal or external. When a manufacturing company cannot produce enough of its products to meet market demands, even though enough raw materials are available, then internal constraints are the most likely cause. On the other hand, if the company has enough products to meet market demand, but cannot sell all that it manufactures then that may be due to external constraints, such as fierce competition, market saturation, or economic recession.

Though every system has its limitations, constraint analysis can identify those that have the greatest impact. By focusing on those one may be able to improve a system's performance. For example, in a chemical plant, improving flow of raw materials may have the greatest impact on its productivity. In that case, the flow of a raw material may be improved with comparative ease by replacing the feed pump with one that can handle higher flows.

In a piece-part manufacturing plant, the constraint is often the availability of certain machines or processing units. However, for a company that employs a large number of skilled knowledge workers, the constraint could be the availability of a few key employees.

For a person with limited family income, constraints can come in many forms, such as lack of money, lack of good health, lack of opportunities, and deficiencies in education and training. However, psychological barriers pose even greater constraints in our lives. They could be lack of self-will, lack of self-worth, hopelessness, and depression.

6.3.2 The Case of Heidi

Heidi is in sixth grade, she is intelligent and had been doing well academically. However, for the last few months, her teacher noticed her lack of motivation. She was not finishing her assignments on time and she looked withdrawn with diminished attention span in classroom. The teacher first tried to push her to be more attentive and finish assignments on time, but that did not work and her academic grades continued to fall. Her case was referred to the head teacher, who called her mother for a discussion in order to understand Heidi's situation at home. Heidi's mother is a young widow, who until recently paid a lot of attention to her and to her studies. However, for the last few months, she has been busy courting a young man, which has led to emotional neglect of her children, though she continued to provide their material needs. By being perceptive, the head teacher was able to understand the situation and to point out to Heidi's mother the need for emotional support to her daughter. Heidi's mother promised to spend more quality time with her children and take more interest in their studies. To a great relief to Heidi's teachers, within a few weeks, Heidi's academic performance began to improve. Here, the perceived lack of mother's love and encouragement proved to be a major constraint in the psychological well-being of Heidi.

Thus, the performance of a student may be improved by first finding the constraints that are preventing her from being a high achiever rather than pushing her to spend more time in studies. Contrary to a common perception like lack of attention or motivation, the barriers may actually range from inadequate parental support, a home environment not conducive to study, to poor nutrition.

6.4 Biological Systems

Nature is a superb optimizer, where natural systems often exhibit minimal use of energy and other resources. Trees have naturally evolved so that their leaves are arranged in such a way as to absorb maximum amount of sunlight. A houseplant placed near a window, if left undisturbed, bends towards the window to get more sun. Photosynthesis itself is a very good example of nature's efficiency, where water and carbon dioxide gas combine in presence of sunlight to produce sugar, cellulose, and lignin that are needed for a plant to grow and survive.

Thousands of years of evolution has eliminated those species that were less efficient leading to the survival of plants and animals that are better optimized to survive in their immediate environments. Some may argue that nature is also wasteful. An oak tree produces thousands of acorns per year, and most of those do not turn into viable trees. Some are eaten by squirrels, some germinate, but most of the saplings do not survive. Only a miniscule number grow to become viable oak trees, which seems quite wasteful. However, from an evolutionary point of view, the primary purpose of a species is its survival. For an oak tree, the purpose of maximizing the chances of survival of its offspring is enhanced by the production of such a large number of acorns.

All natural systems have balanced or optimized growth rates. A system often gets unbalanced with serious implications when the growth rate of one of the components or variables is pushed to a higher gear, while others stay put. An example of that is cancer in which groups of cells display uncontrolled growth through division beyond normal limits.

6.5 Industrial Systems

Most industrial systems need to be optimized to maintain and improve their productivity and profitability. That requires improvements in product quality and reduction of production cost by minimizing the use of raw materials and energy, and lessening wastage. Optimization of a chemical process, such as production of polyvinyl chloride (PVC), a commonly used plastic, or distillation of crude oil in a refinery involves a large number of interdependent process variables.

Distillation of crude oil is a method for separating lighter components, such as gasoline from the heavier ones, such as jet fuel, diesel oil, heating oil, and asphalt. It is carried out in a tower where heated crude oil is pumped near its middle (Fig. 6.3). By controlling flow of hot oil from the reboiler at the bottom of the tower and altering flow of cold reflux at the top, the lower part of the tower is maintained at a much higher temperature than at the top. Inside the tower are a large number of trays where lighter components of the crude migrate upwards while heavier ones flow down. To get the desired level of separation, the in-flow of crude oil, the temperatures at the top and bottom of the tower, and flow rate of the distillates are closely monitored and controlled. Thus, the control system is made up of several control loops. These loops adjust the process variables as needed to compensate for the changes due to disturbances during plant operation. However, these control loops affect one another. A change in one variable, such as the quality of crude oil, requires changes in other process variables.

A set of simple control loops can keep the operation of crude oil distillation in a stable state, but cannot optimize it for changes in the operating conditions, such as change in crude oil composition or its input flow rate. It would require frequent interventions of the operators to readjust set points of a number of control loops, but that may still produce uncertain results. Thus, the optimization of a distillation process requires a unified approach in monitoring and control of a number of variables in the tower, which is termed as multivariable control.

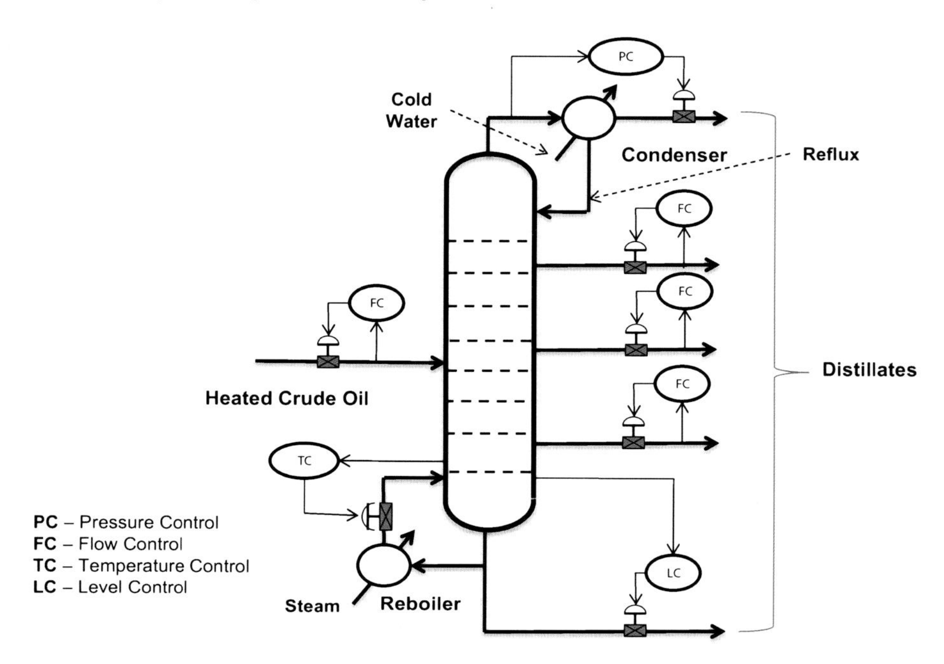

Fig. 6.3 A distillation column

6.5.1 Multivariable Control

Most of the discussions on controls so far have assumed that the system variables can be controlled independently. However, in real world, this is not often the case. Sometimes, controlling the loops independently causes a little problem, but often it is necessary to consider these dependencies to optimize the performance. Distillation column is an example of a multivariable system, but there are many others, such as power plants, robots, and national economy.

For example, an airline company's profit may be affected by fare price, load factors, fuel price, advertising expenses, passage amenities, and employee salaries. Here, fare price, load factor, and fuel price may be considered as independent variables, while profit, advertising expenses, employee salaries, and passenger amenities may be considered as dependent variables. Any change in one of the independent variables may result in a change in one or more dependent variables and in profitability.[1]

[1] The categorization of independent and dependent variables is not always rigid; it depends on the circumstances and the way a system is perceived.

Thus, an increase in fuel price—an independent variable, will decrease profitability, which may require curtailing one or more dependent variables, such as advertisement budget and passenger amenities. In a competitive situation, a reduction in fare will increase load factor, which may increase profitability in the short run. However, if other airlines make matching price reductions, then the increase in revenues will be nullified and the company may end up with a net decrease in profitability. This is an example of interrelationships of variables in a business system.

6.5.2 A Fruit Juice Blending System

A fruit juice blending system (Fig. 6.4), whose conceptual model was described in an earlier chapter, blends apple and grape juice in a precise ratio. Adjusting either of the inlet flows *F1* or *F2* will affect the composition. Manipulating both the valves simultaneously to maintain grade can be difficult and confusing, so the simplest method would be to keep one of the flows constant, while adjusting other to maintain the desired product composition. This approach will work well in many situations, but will not be appropriate if a constant flow of blended product is required for downstream packaging system.

Performance of the system is improved by controlling *F2* by the flow controller for the blended product and by controlling *F1* by a ratio controller (RC). At the start of a run, the ratio is set to a nominal value, which is adjusted (fine-tuned), by the composition controller as the blending progresses

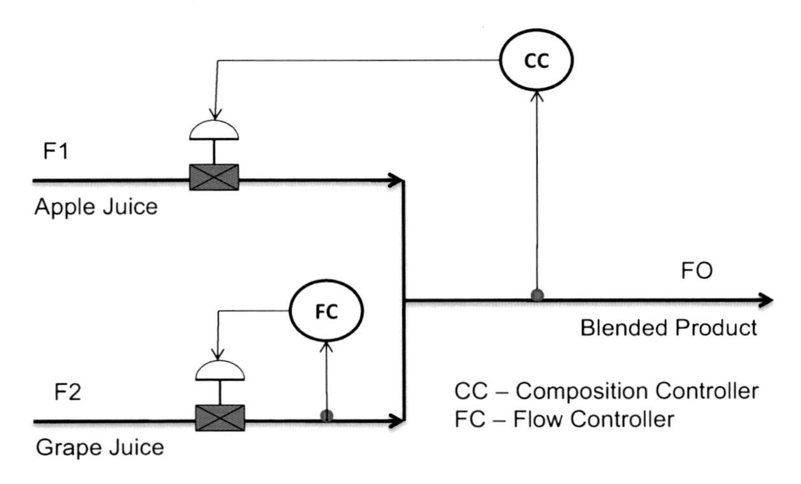

Fig. 6.4 A simple blending control system

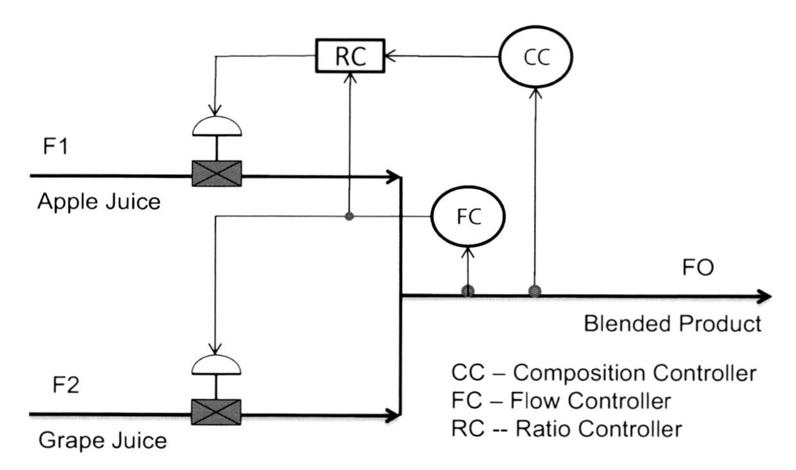

Fig. 6.5 An improved blending control system

(Fig. 6.5). This is an example where two independent variables may be controlled by a cascaded loop to produce the desired result.

There are various techniques for optimizing multivariable processes, such as model-based control and hill climbing.[2]

6.5.3 Model-Based Control

In a simple feedback loop, the controller does not use any model; its output depends on the error between the set point or objective and the feedback of measured value. There the integral control function takes into account the total accumulated error, and derivative control function takes action based on the rate of change of error. Such a controller has a very limited memory and little anticipatory function and performs poorly when attempting to control processes with large lags and multiple interacting variables.

These limitations of simple feedback control have led to the design of feed forward and model-based control. When driving a car, for example, I may push the gas pedal when I am approaching a hill rather than wait until the car starts to slowdown, thus, trying to maintain an even speed. In the case of crude oil distillation, I may set up a strategy for adjusting the temperature profile of the column based on the viscosity of the crude oil rather than

[2] There are a number of books available that deal with these problems in more details. Interested readers will find some of them in the Bibliography.

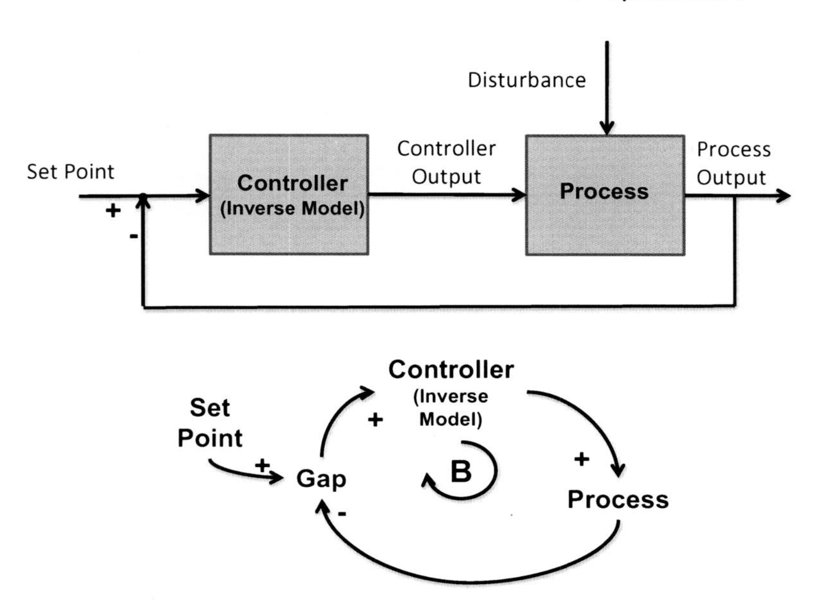

Fig. 6.6 Model-based control

waiting until the composition of the products at the outlet start to change. These types of actions that are based on anticipated changes are termed as feed forward control, which is mostly used in conjunction with feedback control. For right applications, they offer better control over simple feedback loops.

In a model-based control (MBC), we go a step further by making control decisions based on the model of a system. That, however, requires an accurate model of the system to be controlled. The model may be based either on the detailed knowledge of its inner workings (white box model) or based on its external behavior patterns (black box model).

A common way to control a process using a model is to use an inverse model as the controller (Fig. 6.6). An inverse model, as the name implies, is the reverse of a normal model that mimics a system. Such a model generates the necessary inputs to a process when fed with the desired output values (set points). However, as models are never perfect, this ideal situation is impossible to achieve except for very simple systems. Thus, there is always a mismatch between the process and its model, which needs to be corrected while running such a system. That is usually accomplished by putting a normal model in parallel to the process (Fig. 6.7). In this arrangement, the difference between the system and model outputs represents the modeling error. This difference is used by the controller to compensate the effects of mismatch.

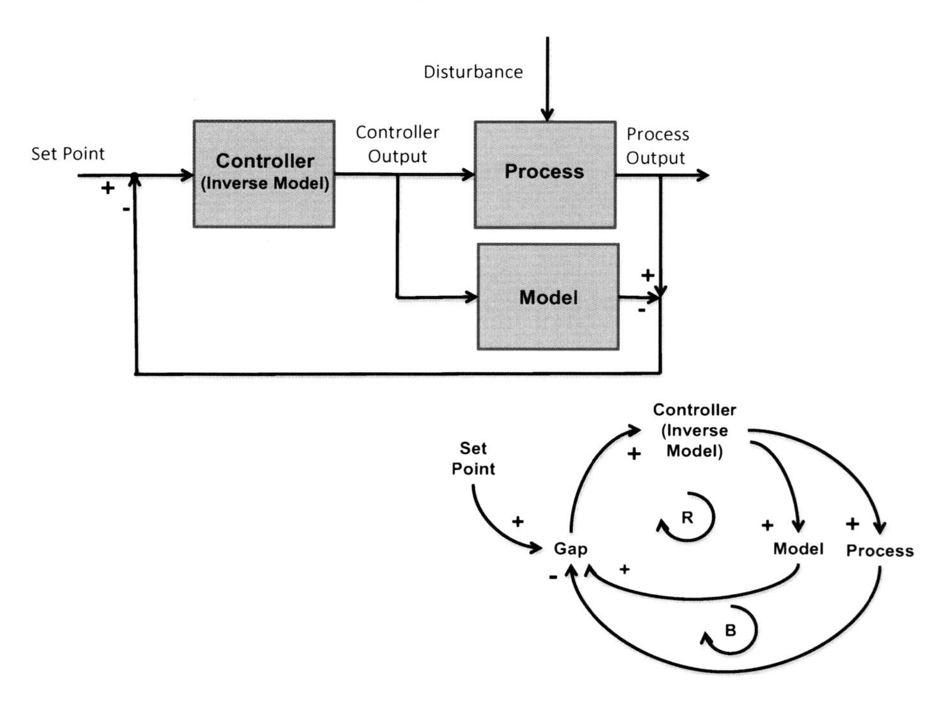

Fig. 6.7 A generic model-based control system

Model-based control is used extensively in industry where multiple process inputs and outputs interact with each other, and there are long delays between control actions and resulting change in process outputs. MBC provides improved control for such a situation by anticipating and compensating for interactions between variables and control loops. In other words, the predictive nature of the process model allows control loops to minimize the detrimental influence of a disturbance even before the actual effect of the disturbance is felt on the loop. In addition, MBC allows optimization of variables within the imposed constraints.

Originally developed to meet specialized control needs of power plants and petroleum refineries, MBC technology can now be found in a wide variety of application areas including chemicals, food processing, automotive, and aerospace applications.

MBC is well suited for optimization and efficient running of a system whose dynamics is well understood, that is, it can be modeled with reasonable accuracy. For many processes, such as chemical plants, crude oil distillation, or drug production, accurate process models based on first principles may be difficult to generate. This is especially true for plants that are already in existence. In those cases, black box or empirical models may be used.

6.5.4 Hill Climbing

Let us suppose that a person is at the foot of a hill on a misty day when the peak and a large part of the hill are not visible (Fig. 6.8). The climber takes a small step forward, backward, or sideways to find whether it takes him up or down. Then, he takes the direction, which takes him upwards. He continues repeating this process at every step until he finds that he cannot go up any further in any direction. He then concludes that he has reached the peak.

Hill climbing is a computer algorithm or procedure that optimizes a multivariable system by mimicking the hill climber. The procedure makes a small change to a controlled variable and notes the result. If it is in the positive direction, then the process is repeated. However, if that is negative, then the program goes back to the previous position, makes a small change to another variable, and repeats the process. This is analogous to a step a climber takes to get towards the top of a hill. The process is continued by making incremental changes to all independent variables until there is no further progress.

Hill climbing is good for finding a local optimum as in the case of a single continuous hilly structure in a region. However, there is no guarantee that it will be able to find the best possible solution when there are a number of hills and valleys in the region (Fig. 6.9). In fact, the process can get stuck in a valley between two hills or in other areas where the topography is flat.

Where finding global optimum is the aim, a hill climbing procedure needs to be continued by accepting sideways and or even downhill moves, while noting local optima, in order to avoid repeated visits to the same place. Thus,

Fig. 6.8 Hill climbing optimization

Fig. 6.9 Hill climbing optimization with multiple peaks

a global optimal solution may be obtained after repeated attempts to find local optima, but there is no guarantee that the global optimum will be found.

It is also possible to put multiple "climbers" at different places on the surface on the basis that one of them is likely to find the global optimum. Alternately, one can run a number of experiments with the same "climber," starting at a different random point each time and noting optimum values to find the global optimum.

Hill climbing method has been used in many industrial systems, where multiple variables are manipulated to get an optimum result. In a chemical process, they may be temperature, pressure, and rate of flow of raw materials. In such a case, each one of these variables may be changed in small increments while the resultant product qualities are noted. An incremental change is reversed when the quality is found to deteriorate, otherwise further incremental changes are made, usually one variable at a time, until an optimum is reached. Hill climbing method may be applied directly in a process to optimize it in real time or it can be tried on a model of the process.

Such iterative processes are common in real life. A child trying to influence her mother to buy a toy for her may try different tact when her first request is refused. She may continue modifying her approach until her objective is realized. Similarly, an effective salesperson may try to offer various little incentives to a prospective customer to close a deal.

6.6 Kaizen: Optimization in Small Steps

Kaizen comes from two words in Japanese language, where kai means change and Zen means better. Kaizen now means a process for continuous improvement in small steps. In Japan, it is considered a way of life for most individuals, businesses, and social organizations. Kaizen is like a hill climbing procedure, where both large and small problems are solved or a system optimized by multiple small steps in the right direction. For an individual, kaizen can bring benefits in professional, social, and domestic lives. Using the same principles, manufacturing and business organizations can also dramatically improve their performance.

Innovation-centric western management philosophy and strategies rely on major advancements and dramatic changes rather than small incremental improvements. That often leads to radical changes and possible unintended consequences. The kaizen way, on the other hand, is often based on common sense and low-cost approach, which allows incremental progress. It is also a low risk approach, because it allows stepping back when an incremental step does not work well. Additionally, small steps circumvent the brain's built-in resistance to new behaviors, thus, to bring in great changes it is easier to start small. The philosophy of kaizen:

- Ask small questions
- Think small thoughts
- Take small actions
- Solve small problems

In real life, we tend to perceive that major changes come only through big thoughts, big actions, and by having the capacity to solve big problems. We are so used to living with minor irritants that it is not always easy to identify them, let alone make corrections. However, these irritations have a way of gaining inertia that blocks the paths to change. By training to spot and solve small problems, as they occur, one can often avoid remedies that are more painful later.

In our daily life, it is possible to train ourselves to see small warning signs more clearly by recalling past mistakes, identifying recent mistakes, finding whether small mistakes have the potential to create larger problems, and trying to solve them in small steps.

6.6.1 Kaizen in Daily Life

When an individual or a family, for example, is trying to reduce energy consumption, it can take many small steps in that direction, such as:

- Switch off the lights in a room when not in use.
- Incrementally replace existing lamp bulbs with those that are more energy efficient.

- Adjust working day to align with the available daylight.
- In winter, put on more warm clothes and reduce the thermostat setting.
- Plug all cracks in door and window joints to minimize cold air getting in.
- Put extra insulation in walls and ceilings.
- Buy more energy efficient appliances, when replacing the existing ones.
- Install an electric meter at an easy to observe location to track the power consumption.
- Buy a smaller car with higher gasoline mileage when replacing the old clunker.
- Use public transport whenever that is convenient.
- Use a bicycle for short commutes.
- Air-dry clothes after washing rather than using an electric clothes dryer.
- Install solar panels to generate electricity and hot water.
- When building a new house, align its windows to face south to get maximum exposure to sunlight.

The list above is neither exhaustive nor prioritized; also, each individual item may not contribute to a significant reduction in energy consumption. However, it indicates a direction where a combination of small steps can lead to a significant amount of savings.

6.6.2 Kaizen in Manufacturing

The kaizen way of improving quality in a manufacturing process is to spot any defect—large or small at the earliest possible stage and fixing both the product and the process at that time. Rather than leaving it until the end of the manufacturing cycle or when the product is in use.

As many large problems have modest beginnings, it is advantageous to fix a problem early when it is easier to fix. The problems between managers and subordinates, supervisors and workers, and customers and suppliers can often be remedied more easily if detected and addressed early. Similar are the situations for machine breakdowns, safety problems, and quality issues.

Kaizen has helped some major companies like Toyota to gain competitive advantage. In order to improve productivity and to solve specific problems, Toyota often sets up a team made up of a cross-section of employees who may never have worked together before. The assembled team of people, with a broad range of job skills, participates in problem-solving sessions, where they are expected to come up with solutions for improvement and ideas for implementing them to get immediate benefits. The team's task is to review a particular process, look at each step, pinpoint any problems, and then brainstorm solutions. Such teams have greatly benefited Toyota in improving quality, reducing waste, and in enhancing productivity.

Small steps of kaizen and giant leaps of innovation are not mutually exclusive; together they are formidable in solving problems and in optimizing systems.

6.7 An Objective Way to Optimize a Decision-Making Process

John was entrusted by his employer to select a fleet of automobiles for use by the field sales force in his company. He was asked to select a make and a model that would be most economical to buy and to operate and would be comfortable enough for a long distance travel for the sales personnel. The purchase price was set not to exceed US$ 25,000 per vehicle. John very much liked the car he owned, as he had a good experience with that for a number of years. However, he did not want to be swayed by his personal likes or dislikes in making the selection for his company. In other words, he needed to be objective in making the recommendation.

The task was not that straightforward, because to make the right decision, John needed to optimize a number of criteria or objectives. That included high reliability, good driving satisfaction, and low lifecycle cost. There was, of course, no car in the market that had the highest ratings in all these three areas. Therefore, in making the right decision, he needed to make some tradeoffs.

John decided to create an objective evaluation matrix to help in the analysis and decision-making process (Table 6.1). In order to make sure that he included the viewpoints of a diverse range of people, John set up a committee, which included a number of field sales personnel who were going to use those cars and the controller of the company who was responsible for financing the purchases.

Before the first committee meeting, he created the evaluation matrix on a spreadsheet. He listed three major criteria for choosing a car, which were reliability, driving satisfaction, and lifecycle cost. He listed them in the first column (A) of the spreadsheet. Then he set up lists of sub-criteria for each of these three major criterions, which included items, such as power train, brakes, and body integrity under reliability; driver's comfort and climate control system under driving satisfaction; and purchase price and cost of maintenance under lifecycle cost. He listed them in the second column (B) under each respective major criterion. He then presented that to the committee and noted their comments to create the final version of the list.

Table 6.1 An objective evaluation matrix

Automobile evaluation matrix		Percent weight (%)		Factored percent weight	Automobile ratings			Automobile scores		
Primary criteria	Sub-criteria	Primary criteria	Sub-criteria	weight (%)	No. 1	No. 2	No. 3	No. 1	No. 2	No. 3
A	B	C	D	E	F	G	H	I	J	K
Reliability		35								
	Power train		30	10.50	8	9	7	0.840	0.945	0.735
	Electrical		10	3.50	7	8	6	0.245	0.280	0.210
	Climate system		10	3.50	9	7	8	0.315	0.245	0.280
	Suspension		10	3.50	7	6	7	0.245	0.210	0.245
	Brakes		20	7.00	8	9	7	0.560	0.630	0.490
	Exhaust		5	1.75	5	8	6	0.088	0.140	0.105
	Body integrity		15	5.25	5	7	5	0.263	0.368	0.263
	Total		*100*							
Driving satisfaction		25								
	Driver's comfort		45	11.25	9	8	9	1.013	0.900	1.013
	Climate system		35	8.75	8	8	9	0.700	0.700	0.788
	Audio system		20	5.00	6	6	7	0.300	0.300	0.350
	Total		*100*							
Life cycle cost		40								
	Purchase price after Discounts		40	16.00	9	9.5	9	1.440	1.520	1.440
	Fuel consumption MPG		25	10.00	7	8	7	0.700	0.800	0.700
	Cost of regular maintenance		15	6.00	8	8.5	7.5	0.480	0.510	0.450
	Resale value after 3 years		20	8.00	5	8	5	0.400	0.640	0.400
	Total		*100*							
Total		*100*						*7.588*	*8.188*	*7.468*

Factored percent weight (E) = Percent weight criteria (C) × Percent weight sub-criteria (D)
Automobile score (I, J, and K) = Factored percent weight (E) × Automobile ratings (F, G, and H)
MPG miles per gallon

The next step for John was to get consensus on the weight or the level of importance for each criterion at primary and secondary levels. These weights were specified as percentage, which added up to 100 for a set of criteria. For example, at primary level, reliability, driving satisfaction, and lifecycle cost were assigned 35, 25, and 40 percentage points, respectively (column C). Then, the sub-criteria were assigned their weight factors (column D), such that they added up to 100 % for each of these primary criterion. Assigning these weights took a considerable time and required multiple meetings as there were many divergent views within the committee on their relative importance. John then created a column for factored percentage weight (E), which was a product of the weights of primary and secondary criteria.

Actual rating of the automobile models started only after all the criteria for selection and their weights were finalized. Rather than performing detailed evaluation of all the cars that were in the market, which would be very time consuming, the committee decided to eliminate those that were less likely to meet their requirements. That left them with three most promising models for detailed evaluation.

These three models were evaluated in the scale of 0 to 10 for each sub-criterion. A zero rating was given when a function was found to be not available and ten when it fully met the requirements. The spreadsheet formulas automatically multiplied the ratings with the factored weights to generate the score for each sub-criterion and the total score for each of the three models. It was then easy to choose the model that scored the highest point.

6.7.1 Steps for an Objective Decision-Making Process

1. Assemble a committee of knowledgeable persons of diverse backgrounds.
2. Decide on main criteria for decision-making.
3. Decide on sub-criteria under each main criterion.
4. Set additional levels of sub-criteria if needed.
5. Set "weights" for each main criterion and sub-criterion as a percentage, where the total for all criteria or for each set of sub-criteria adds to 100 %.
6. Generate factored percentage weights for each lowest level sub-criterion by multiplying respective percentage weights for main criterion with the sub-criterion.
7. Rate each of the objects under consideration (automobile model in this case) in a scale of 0 to 10 for each lowest level sub-criterion.
8. Multiply each rating number with its weight to generate the score for each object under consideration.
9. Add all ratings numbers for an object to generate the total score and select one that gets the highest score.

The method outlined here is an objective way to optimize a decision-making process. This procedure can be applied in many different situations, where multiple criteria need to be considered. The main difficulty in following this process lies in generating an unbiased set of factored weights for the list of criteria or objectives. The bias is minimized when these weights are agreed by a committee, rather than by an individual.

This method, however, may not work well in a situation where an individual has to make a decision to buy an article, such as an automobile, for his or her personal use. There, objective decision-making is often difficult as personal emotions play a major part. Sales persons for high-value items such as automobiles and jewelries often exploit these emotions by inducing the prospective customers to fall in love with the object they may buy before discussing price, thus reducing objectivity even further.

6.8 Optimization: Challenges and Opportunities

In our daily life, we interact with numerous mechanical, social, cultural, and institutional systems that are around us. We would like to optimize them if we can and also optimize our interactions with them as much as possible. Parents want their children to learn at school in an optimized fashion while students often want to optimize their study habits. From tycoons to small business owners, all want their businesses to run in an optimized fashion. Global warming and the clamor for green technology are pushing us towards greater optimization of systems to minimize energy use and reduce emissions and wastes.

Scientists, engineers, and technicians who design, build, and operate utilities such as electric power generation and distribution, water supply, and communication systems spend a lot of time in optimizing their systems. So are the manufacturers of various systems that consumers use. Additionally, consumers who interact and use these systems want their interactions to be as efficient and optimized as possible.

List of things that need to be optimized is endless, and so are the many different strategies for achieving them. However, there are many commonalities in optimization processes, which become apparent when we start looking at them as systems. Scientists and engineers often try to emulate common sense and nature's ways of optimization in designing and operating electromechanical and industrial systems. Similarly, social scientists and general-public can learn many strategies and methods from natural scientists and engineers and apply them in social and other systems, and in everyday life situations.

Following are some of the major barriers that need to be taken into account in optimizing a system.

Mental Inertia (Deeply Ingrained Assumptions and Beliefs)

Mental inertia often limits our capability to change ourselves or to change the systems that we work with. There is always a fear factor at the back of our consciousness, which makes us wary of any change. Our beliefs and attitudes are mental models formed early in life, which we need to question as we grow up. We need to take a hard look at those to decide whether they fit with current realities. As political and economic situations alter, it is imperative for us to understand at least the broad implications of such changes to be able to interact with them optimally.

Mental inertia is usually not a major issue for scientists and technologists responsible for designing and working with engineered systems. They tend to be more flexible in their approach towards these systems as that have little to do with their beliefs and mental assumptions. However, they also often tend to stay with what has worked in the past rather than readily accepting new ideas. Additionally, "Not Invented Here" (NIH) can be a powerful drag to innovations in system design and optimization. NIH is a tendency that is present among both individuals and organizations that rejects external solutions in favor of internally developed ones. Generally, smaller companies suffer less from NIH than larger corporations. Similarly, nations and societies that are in rapid development are more willing to embrace external ideas than those that are already in a more developed state.

Working with Imperfect or Outdated Models

In industrial control, the model of a process is often used to predict its behavior and to take corrective actions before a problem becomes apparent. However, a model is never perfect and often a system is time variant. Therefore, the model needs to be updated at frequent intervals to represent the actual system more accurately.

We use mental models to guide many of our actions, be that in our home, at our work, or in social situations. Most of our mental models are grossly simplified, as our brain cannot deal with a large number of variables and objectives at the same time. Nor can it effectively deal with nonlinearities that are inherent in many systems. One way to deal with this type of situation is to decompose a large system to a hierarchy of models, where each of them may be relatively simple and easy to comprehend (this topic will be dealt in detail in a later chapter).

More often than not, we fail to update our mental models. That has a lot to do with our mental inertia, our deeply held assumptions and beliefs as explained earlier. One needs to be flexible enough to be able to change one's mental model with the change in reality. However, changing a mental model frequently, without any tangible changes in the system under consideration may lead to vacillation and inaction, which needs to be avoided.

Long Feedback Time, Making Consequences for Actions Not Readily Apparent

Long feedback time not only makes a system sluggish and inefficient. It also makes it harder to perceive the effects of an action, making the system more difficult to control. Just reducing delay and dead time, often work wonders. An employee's performance increases rapidly when her manager gives prompt feedback on her performance rather than waiting for the year-end review. A person dealing with a bureaucratic system with long delays may find it advantageous to approach the decision-maker directly rather than taking the usual slow route.

In an engineered system, a long feedback time or dead time plays a significantly negative role in its controllability and optimization. If an automobile has a dead time between the steering wheel and the tires, such that the wheels do not respond for, say, 2 s every time one turns the steering wheel, then that would have a serious negative effect on its drivability. In a chemical reactor, the controllability suffers if the temperature controller does not get any response for a few minutes for increased flow of steam to heat up the reactants. Recommended solution to such problems is to reduce the dead time to a minimum, which may be possible by better system design and better positioning of sensors that provide the feedback. Thus, for a system with long delays, model-based control with feedback often works better than only feedback control.

It is, however, very important to recognize those delays that are inherent in a system, which cannot be minimized. There, a change made today may take days, months, or years to show tangible effects. A change in the educational system, for example, may show its effect years later, when the students graduate or even later when they are working adults. An economic stimulus may take years to show its full effects. Similarly, a bad policy decision can show its detrimental effects years after a politician who made the decision has retired from office. Additionally, an action may lead to oscillation in a system, where the initial result may look good, but later results may not be so benign. In those cases, it is imperative for us to understand and appreciate the nature and period of these delays and refrain from making rash judgments.

Inability to Overcome Constraints

In many instances, the main problem in optimizing a system is the identification of constraints that lie hidden. Once a constraint is identified, one can formulate strategies to eliminate or minimize it or to work around it. For hard constraints, work around is the only course of action, but for soft constraints one can attempt to eliminate or minimize them. For most systems, it is a better strategy to tackle the constraints first rather than pushing the systems to their limits. For example, pushing a sales person to generate greater sales may not be as effective as removing the constraints he or she may be facing, such as lack of effective sales literatures or adequate backup support from the engineering or design department. In a manufacturing environment, identifying and reducing bottlenecks may help significantly in gaining higher production rates. Thus, a good manager is often a facilitator who identifies and minimizes the constraints faced by her underlings.

Pushing a System beyond Its Optimal Growth Rate

Optimal growth may be viewed as a balanced growth, where the growth of one subsystem does not lead to a severe misalignment with other subsystems or with other systems that are strongly coupled. For example, making it easier to own motor vehicles without corresponding improvements in road systems leads to an increased congestion. Similar is the case with cities that grow too fast without corresponding growth of their infrastructures.

When a business organization grows too fast, it often fails, not because of its growth rate per se, but because of the system getting out of balance. The imbalance is caused when there is no corresponding growth of other activities, such as increased management responsibilities, uniform quality control, increased after-sale service, and more capital expenditures for new facilities.

7

Distributed Intelligence

Abstract Natural living systems, such as plants and trees, have decentralized structure. A tree survives even when some of its branches are chopped off. Grass continues to grow after repeated mowing. However, top-down command and control is the norm for engineered systems. There, a decision made at the highest level is transmitted to lower levels, where the task is usually executed in a very predictable fashion. Now with increasing size and complexity of engineered and other systems, there is a growing realization that their proper design and their needed robustness can only be achieved by increasing decentralization of their functions and their local intelligence. That is leading to the development of intelligent agent-based systems, which are applicable to wide varieties of applications.

This chapter starts with the description of an on-board chilled water supply system on a navy ship whose robustness was increased by decentralizing its control structure. Then, there is discussion on the need for decentralized structure for complex systems. That is followed by a narrative on how computer software has evolved from simple programs with subroutines, to objects, and intelligent agents. Then, the advantages of intelligent agents are discussed in detail with application examples. Finally, there is a short discussion on agent-based modeling and simulation.

A. Ghosh, *Dynamic Systems for Everyone*,
DOI 10.1007/978-3-319-43943-3_7

7.1 A Robust Chilled Water System

In the early 1990s, the Office of Naval Research in the USA was looking for a highly survivable robust system for distribution of chilled water in some of its ships. The supply of chilled water is vital to the proper working of an array of on-board equipment, such as combat systems, communication systems, radar, and sonar. The chilled water system removed heat generated by such equipment and was essential for their proper functioning. An interruption of its supply, even for a few minutes, could cause some of the equipment to shut down or fail to operate properly.

The major requirement was that the chilled water system must continue to operate even after a major disturbance, such as a fire or missile strike somewhere on a ship. The resultant goal was for a highly resilient system with no single point of failure. In a conventional chilled water system, one controller is quite capable of handling its normal operations. However, a single controller makes the system vulnerable to a single point of failure. A single controller also makes the system rigid and difficult to expand and maintain as its requirements evolve. After investigating many different approaches to address this problem, a system with highly distributed intelligence was proposed.

The system was then designed and built with a number of intelligent controllers that offered both reasoning and real-time control functions. Each of these controllers was physically located near a set of equipment that needed chilled water to form a node (Fig. 7.1). They took actions based on local needs while being aware of the conditions of other such nodes. Additionally, these nodes included diagnostics and system reconfiguration function for taking care of failure conditions.

In effect, intelligence in each node in the distributed system decided what to do when a portion of the system is no longer operational. Thus, the system was able to perform automatic reconfiguration of water distribution system when a part of it was inoperable.

In a centralized system, a control engineer needs to visualize and then design for every possible combination of failures ahead of time and the failure of a single controller could still shut down the whole system. In a system with distributed intelligence, there are more controllers but each controller needs much simpler logic to define its actions in case of failure of another node. Thus, a highly distributed intelligent system provides higher level of resilience, more flexibility for expansion, and a greater ease of maintenance.

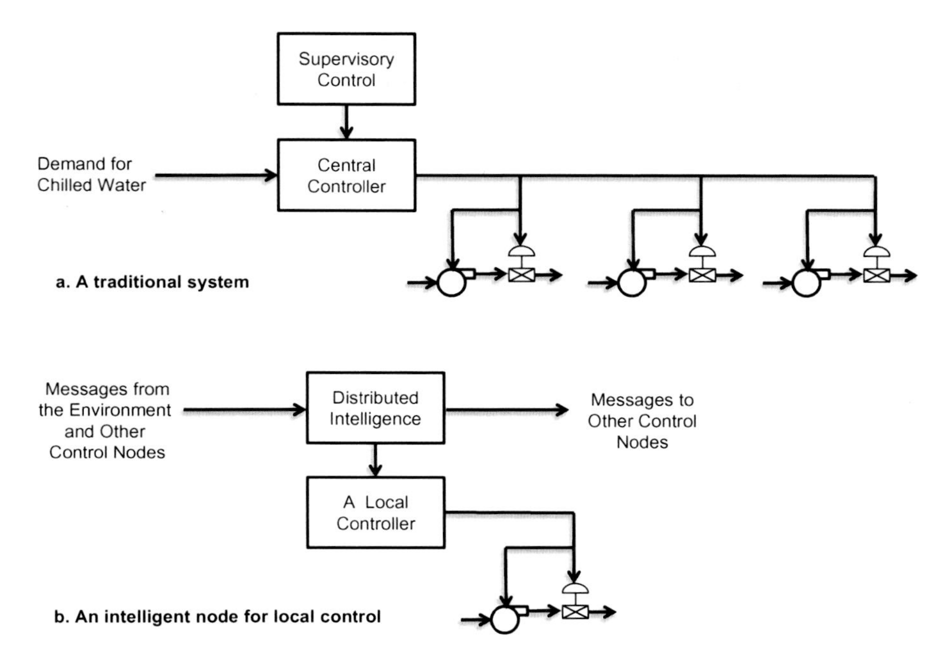

Fig. 7.1 A chilled water control system

7.2 Need for a Decentralized Structure

Natural living systems, such as plants and trees, have a decentralized structure. A tree survives even when some of its branches are chopped off. Grass continues to grow after repeated mowing. Biological systems, such as animals and human beings, also have decentralized structures, perhaps to a lesser extent than plants and trees. Even though a brain directs all major decision-making processes in an animal, its vital organs work independently. Normally, each individual cell in the body of a human being or an animal works independently, performing its tasks in an efficient manner. For example, when a person or an animal suffers a minor wound, local skin and blood cells take appropriate actions to reduce bleeding and chances of infection without the conscious decision from the brain.

However, engineered systems are generally structured with top-down command and control. A decision made at the highest level is transmitted to lower levels, where the directed task is normally executed in a very predictable and exact way. The expectation of predictability is not surprising. Human beings design and build mechanical and electromechanical systems that are extensions of their normal capabilities but are supposed to perform their functions in exact manner. These systems may be complex, such as automobiles, airplanes, elevators, domestic electric supply systems, manufacturing plants, and the like, but all their functions are predesigned.

It is amazing that complex systems, like these, work so reliably most of the time. However, with increased capabilities of man-made systems there are added complexities, which make them more difficult to design, build, and maintain. Now, there is a growing realization on the limitations in human comprehension and capabilities in design, maintenance, and proper working of these complex systems. Interconnected power grids is a case to point, where the interconnections have led to increased reliability of power distribution in geographically large areas but that has also increased the propagation of faults. The massive and widespread power outage that occurred in parts of Northeastern and Midwestern USA and parts of Canada in 2003 is an example of such vulnerability.

There are various definitions of complexity but they usually are associated with a large number of subsystems that are of different types and exhibit nonlinear behavior. Complexity does not always increase with the size of a system; a large static system may be less complex than a smaller system, which demands more flexibility and adaptations. For example, the road system in a city may be complex but it can be mapped and it is unlikely to change significantly over time. However, the market for small cars is subjected to change in time, based on several factors, such as the state of economy, gasoline price, and demography, making it more complex and difficult to predict.

Moreover, when a dynamic complex system fails, detection of its cause and capabilities needed to fix it may turn out to be a major problem. First, detection of the cause of failure may be difficult because of its dynamic nature. Its operating conditions immediately before its failure may hold the key but that is not always easy to discern. Then, it may be quite difficult to fix it unless the system has been designed in a modular fashion.

Resilience of a system in case of failure of one or more of its subsystems is another important issue. A system has little resilience if the failure of a subsystem makes the whole system inoperative. Ideally, a system should have sufficient backup so that a single failure will not affect its workings. The next best situation is where a single failure leads to a partial shutdown, while rest of the system continues to work properly. These demands are leading to the development of highly distributed systems with significant intelligence at lower functional levels. Examples of resilient systems include long-distance telephone circuits and Internet. They remain functional even when a part of a system is broken by being able to reroute information and messages around the damaged areas.

7.3 Swarm Intelligence

Social insects, such as bees and ants, show a strong sense of group behavior. That includes self-organization, flexibility, and robustness along with collective problem-solving capabilities. Instead of centralized and rigid control structures, the behavior of these social groups is characterized by decentralized and parallel execution of activities. Even though each insect has limited intelligence and capabilities, they closely cooperate to solve complex problems, exhibiting abilities to adapt without external intervention.

A colony of ants, for example, is able to optimize its foraging efforts, where each ant follows a set of simple rules. When an ant walks, it drops pheromone on its path; however, the strength of this chemical marker dissipates with time. Thus, when it finds a food source, it carries a small part of the food back to the nest, dropping more pheromones on the way. When other ants get the smell of the chemical, they follow the markers to get toward that food source. Initially, ants may follow a number of routes but the ant that happens to take the shortest route return to the nest more quickly thus keeping more pheromones active than those following longer paths. Subsequently, more ants take the shortest route as that has the highest level of chemical markers. As more ants follow that path they continue to reinforce the chemical trail, thus keeping food gathering process optimized. Pheromones, however, evaporate with time so the ants have to start all over again each morning, leading to the possibility of finding new and alternate food source.

This example of ants' self-organization does not follow any top-down or complex set of procedures. It follows simple rules to optimize the behavior of a large system. It shows an aspect of self-organization, where a system is able to adapt its behavior dynamically to changing conditions without external or supervisory intervention. Such a system is also quite robust as there is little adverse effect on the system when an individual insect is incapacitated or dies in action. Nature offers numerous other examples of such intelligence, such as hive building of honeybees, mound building by certain species of termites, or collective hunting patterns of wolves and wild dogs.

Thus, swarm intelligence is the property of a system where relatively unsophisticated collective behaviors of a set of subsystems interacting locally with their environment cause consistent functional patterns to emerge. It provides a basis for exploring collective and distributed problem solving with less or no centralized control.

7.4 Distributed Autonomous Systems

As stated before, conventional mechanical and electromechanical systems traditionally have hierarchical architecture with tightly coupled hardware and software modules that exhibit high predictability but low flexibility and robustness. Taking cues from nature, computer scientists and software engineers are now building systems that are distributed, robust, and autonomous. Building blocks for such systems are commonly called software agents or *intelligent agents*.

Intelligent agent may be defined as a person or an entity that acts or has power or authority to act or represent another in an intelligent manner. When you are looking for a house to buy an intelligent real estate agent, for example, will ascertain your budget, your needs, and your likes and dislikes to show you only those houses that are of interest to you. The agent will weed out many other houses that have little chance of gaining your approval, thus making the process less time consuming and more efficient.

Here, of course, we are discussing agents that are software entities. An example is the approval process for a loan application to buy a house, which would involve many functions, such as verification of income, assets, and debts of the applicant, and assessment of the property value. A conventional software program to automate such multiple tasks will be quite complex and difficult. That is because the software programs for each of these tasks have to communicate with each other closely to decide whether the loan application should be approved or not. The problem becomes simpler and more manageable when software agents are used to perform these individual tasks, where the system provides framework for effective communication between these agents.

An intelligent agent-based system can respond to external disruptions more effectively than traditional systems. Examples include rerouting material movement in response to an equipment failure or changing a production plan or schedule without stopping the process.

The field for agent-based applications is quite diverse, which includes business systems, manufacturing, process control, telecommunication, transportation management, and electronic commerce. Thus, the design and deployment of intelligent agent-based systems require a paradigm shift from the way most systems are designed today.

7.5 Development of Distributed Computation

A short discussion on how computer programming has evolved from simple mathematical calculations to agent-based systems with distributed intelligence is appropriate in this context.

Digital computers grew from being mere calculating engines to information processing and controlling of machines in latter part of last century. Since then, computers have become indispensable tools for driving various manufacturing and other mechanical activities. Today, oil refineries, large chemical plants, large pharmaceutical processes, and food manufacturing facilities, along with mechanical parts manufacturing and assembly plants, depend on computers for their day-to-day running operations. That is also becoming increasingly true for other sophisticated electromechanical devices, such as airplanes and large ocean-going vessels, which cannot run safely and efficiently without their help. Penetration of digital devices is increasing as more and more programmable microchips are being embedded in engineered systems ranging from automobiles to washing machines.

Mainframe computers were expensive and bulky. Demand for cost reduction and greater system reliability led to the design of special purpose computer-based distributed control systems with multiple controllers (Fig. 7.2). In a

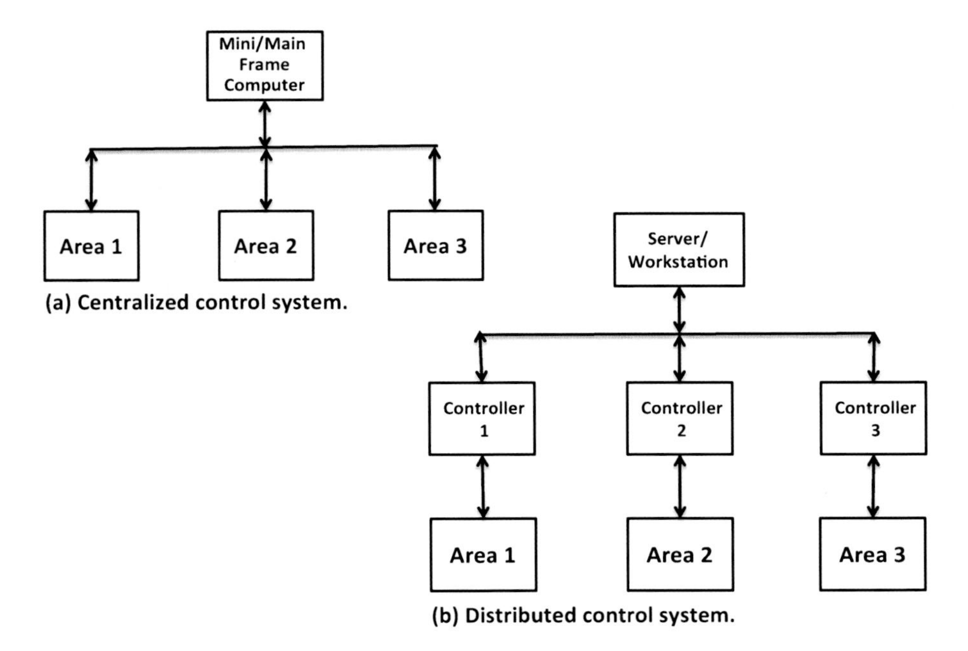

Fig. 7.2 Centralized and distributed control systems

distributed control system, each controller (a digital device) performs real-time control of a subsection of a plant or a system, while minicomputers or personal computers perform supervision and coordination.

Today, a mechanical system, which uses a computer or digital device for its control, cannot be considered in isolation, but needs to be considered as a combination of its mechanical and control parts. Additionally, the functions of a digital device are not merely defined by the microchips and electronic circuitry but also by the software programs that go with it. Thus, there is a growing symbiotic relationship between machines, microchips, and computer programs.

7.5.1 From Monolithic Software to Subroutines, Objects, and Intelligent Agents

Software is the driver of a digital device, which may be a large mainframe, a minicomputer, or a microchip. Software modules specify the way a system or a subsystem behaves. A set of such software instructions is called a program. A program may be specified in machine code (a low-level set of instructions meaningful only to a computer) or in a higher-level language, such as BASIC, FORTRAN, Java, or in some tabular or even graphical language that can be understood by both computers and human beings.

7.5.1.1 Subroutines

As software became more and more complex with increasing sets of instructions, ways were found to modularize them. Where, a subroutine is a set of software instructions that is used repeatedly to perform a certain common function. There are, for example, subroutines in Windows operating system for retrieving characters in a Word file and sending them to a printer. That subroutine may be called on any number of times to print sets of characters. Similarly, there are subroutines that control the opening or closing of a device external to the computer, such as a valve or an electric motor (Fig. 7.3). A subroutine acts at the behest of higher-level programs, other subroutines, or manual instructions.

A subroutine may need a pertinent set of data for its proper execution. For example, a subroutine that adds a raw material to a tank needs to know the quantity of material to be added. Similarly, a subroutine that controls temperature in a reactor needs to know the desired temperature it needs to maintain. Generally, these values are not a part of software code. They are information

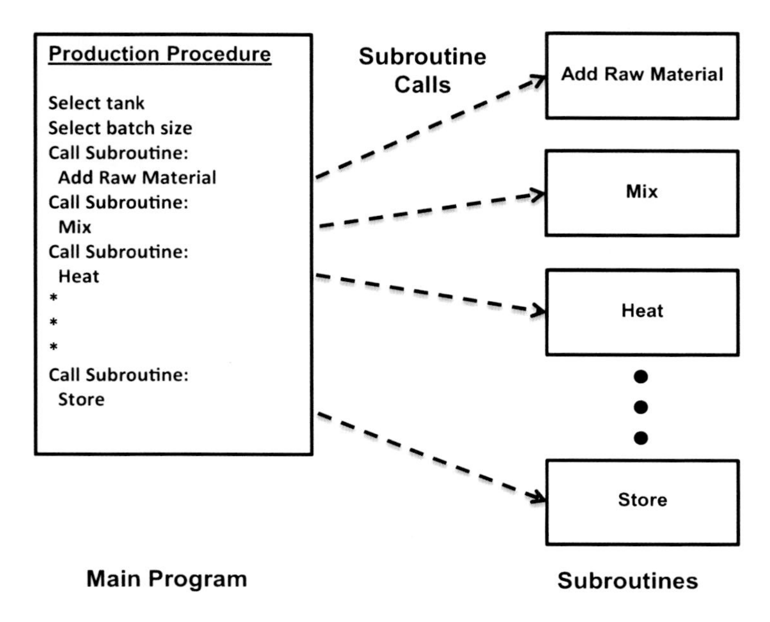

Fig. 7.3 A program with subroutines

that are needed in real time and may be set manually or by other software programs. For that reason, they are usually maintained as separate tables.

7.5.1.2 Objects and Object-Oriented Programming

Object-oriented programming (OOP) takes a more sophisticated approach in designing common functions or subroutines by combining the procedure and data into a single entity called object. OOP is a way of writing a computer program in a modular fashion. These modules are, in most cases, complete programs by themselves.

OOP modules (objects) have the integrity to standalone and decide on their course of actions depending on their own characteristics. They may be compared with human beings who are autonomous, where each person may take a different approach to achieve the same goal. For example, when I invite (in OOP jargon: send message) my friends to a party in my house their mode of transportation may be different. Some may use their car or bicycle, other may use public transport, and those who live nearby may just walk (Fig. 7.4).

An object may interact with other objects to generate specific functions. An object has a firewall around it, that software bugs find difficult to penetrate, making it more robust. Object structure makes it easier to compartmentalize

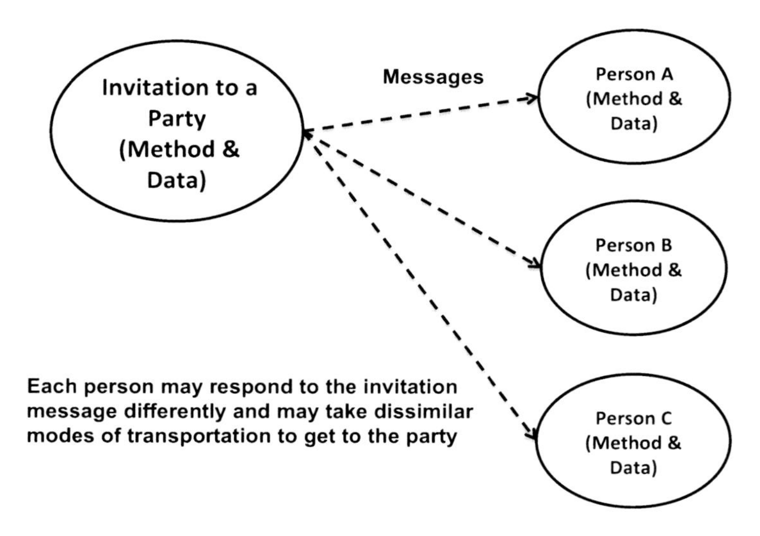

Fig. 7.4 Object interactions

large functions into manageable chunks. One of the goals of OOP is to produce software that is easily modifiable for reuse. Thus, a broken or malfunctioning object may be replaced easily by a healthy one, like replacing a broken part in an automobile. Software developers are now able to buy and sell libraries of such objects, which other developers can use to create large software applications with less effort.

Basic concepts of OOP are encapsulation, inheritance, message passing, and late binding. There is a great deal of freedom in defining an object, which may control a physical entity like a pump or a motor, or a function like calculating monthly profit and loss of a company.

Encapsulation A technique for hiding messy and complex mechanisms inside an object that provides a specific functionality. The mechanisms include data and related procedures, called methods in OOP jargon. For example, from a driver's point of view an automobile's various parts like engine, fuel injection system, suspension, and electrical system are encapsulated. Normally, a driver is not interested in their inner workings as he or she interfaces with them only at a higher level using steering wheel, brake, and gearshift mechanism.

Inheritance This function allows the characteristics of one object to be inherited or copied to another object in the same class. Thus, if there happens to be a number of similar electric motors in an area then, by defining a class called electric motor, all those motors may be controlled by similar pieces of software by copying the functions defined for its class. However, the system has to be flexible enough to allow modification for the needs of individual motors, as necessary. Inheritance is common in real life, where each of us inherit some of the characteristics of our parents and other ancestors, while having some of our own unique individualities.

Message Passing In an object-oriented environment, since the procedures and much of the data are encapsulated, no other object or program can run those procedures directly. They can be initiated only by an object itself. Other objects may request an object to carry out a function by sending a message, then the object at the receiving end may run necessary procedures if that message is accepted as a valid request.

That is like a sales manager asking her sales persons to increase their sales volumes by a certain percentage points in the next quarter without being specific on how to go about doing so. Each sales person may then achieve their part of the overall goal using strategies that are best suited for their individual territory and customer base.

Late Binding Binding is the process of establishing links between objects so that they may communicate as needed. In many conventional systems, these links are established when a system is designed or first set up. This is termed as early binding, which makes the system quite rigid and less flexible. In an object-oriented environment, each object has a unique name, which allows the system to set the link as and when a request for communication is made. This makes the system significantly more flexible allowing addition of new objects or deletion of one or more existing objects as needed.

OOP offers a number of advantages over conventional programming. Encapsulation allows a program to be decomposed into manageable modules whose functions are understood more easily. A system's reliability is enhanced significantly by not allowing direct communications between procedures but only by requests between objects. Inheritance reduces programming efforts by being able to define a type of object only once. Late binding increases flexibility by allowing addition or deletion of objects at any time or moving objects from one processor or computer to another when they are linked to each other.

While objects and the OOP environment offer significant advantages over simple subroutines, an object is usually not a completely independent entity. An object performs its function only when it is activated externally by another object or by a program. Additionally, an object has no learning capability.

7.5.1.3 Development of Intelligent Agents

As stated before, intelligent agents are autonomous, problem-solving computational entities. They are often deployed in environments in which they interact, cooperate, and compete with other agents and/or human beings.

While objects reduce complexity and increase robustness by information hiding and inheritance, they are passive entities that are invoked by external commands. Agents, on the other hand, are autonomous entities capable of exercising choice over their actions and interactions. They are always active, are conscious of their environments, and, thus, are capable of taking autonomous actions based on their objectives (Fig. 7.5). Unlike objects, agents cannot be invoked by other agents or computer programs. An agent acts when it deems necessary based on the information it gathers from its environment. However, the course of actions it takes is based on its internal rules.

A real-life problem requires multiple agents to complete a task. These agents need to interact with one another to achieve their individual or common objectives. These interactions may vary from simple information passing to rich social connections, such as cooperation, coordination, and negotiation. Because agents are flexible problem solvers, operating in an environment over which they have only partial knowledge and control, interactions need to be handled in a similarly flexible manner. Thus, agents have the ability to make context-dependent decisions about the nature and scope of their interactions and to initiate and respond to interactions that were not foreseen at design time.

Adopting an agent-oriented approach requires decomposing a problem into multiple autonomous functions. These functions can then act and interact in flexible ways to achieve their set objectives. Explicit structures and mechanisms are often used to describe and manage the complex and changing web of organizational relationships that exist among agents. Thus, agent-oriented approach avoids the drawbacks of a hierarchical system by distributing the decision-making process to its intelligent parts.

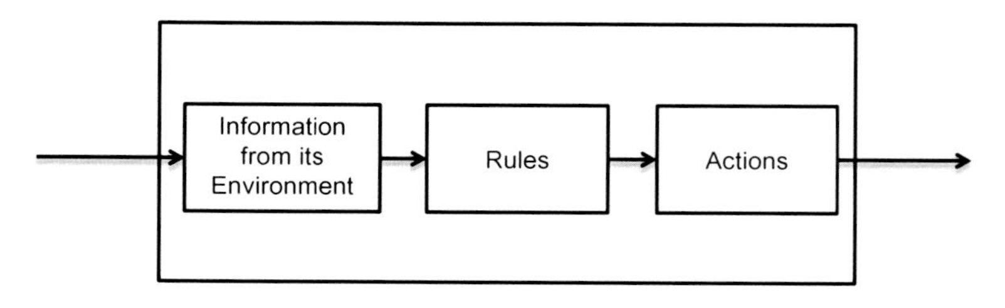

Fig. 7.5 An intelligent agent

7.6 Intelligent Agent Advantage

Agent technology has the potential to affect the lives and works of all of us. As stated before, traditional industrial automation systems have hierarchical architectures with tightly coupled hardware and software modules that have low flexibility and robustness. An agent-based system with higher autonomy and loose couplings between agents promotes increased robustness and flexibility. Thus, intelligent agent-based applications offer increased return on investment where highly distributed control with flexibility and robustness are of paramount importance.

Flexibility and robustness are the main justifications for many agent-based systems. Robustness is achieved mainly because there are less centralized decision-making processes. Additionally, the loss of one of the subsystems does not lead directly to the failure of another.

Agent technology allows reconfiguration of parts of a system without adversely affecting others. It enables reassignments in a production facility, such as adding or removing process equipment or machines on the fly, without reprogramming the entire software system.

An intelligent agent-based system can respond to external disruptions, such as changes in production plan or schedule, more effectively without stopping a process. The system can handle these changes while performing tasks related to equipment changes or failures.

Multi-agent systems (MAS) are groups of agents organized according to specific, precisely defined principles of community, organization, and operation and are supported by an adequate agent platform or infrastructure. MAS offer one of the most promising technologies for making production systems more agile.

Advantages of Agent-Based Systems

- Increased flexibility and robustness
- Ease of modification and reconfiguration
- Ease of redeployment for new applications
- Increased feasibility for highly distributed applications
- Reduction in complexity
- Increased flexibility and autonomy
- Increased decentralization
- Increased intelligence and learning capabilities

7.7 Intelligent Agent Applications

The field for agent-based applications is quite diverse but it is at its infancy. With its technical development and increased ease of use, it will become more widespread in the near future. At the beginning of this chapter, there is the example of intelligent agent application on a chilled water supply system on navy vessels. Following are examples of their use in other areas:

- Real-time control of high-volume, high variety, discrete manufacturing
- Monitoring and control of systems that are highly distributed
- Transportation and material-handling systems
- Production management of frequently disrupted operations
- Supply-chain management
- Online learning and testing of students
- Filtering of college admission applicants
- e-commerce
- Voice recognition
- Robotics

7.7.1 Daimler's Manufacturing Line Control

A typical automotive production or assembly line is a sequential operation where a cell outage can bring an entire line to a standstill. Here, intelligent agents are able to reroute the pieces to an alternate functioning cell, offering fault-tolerant performance.

An intelligent agent application developed by Daimler AG is deployed on one of their production lines in Stuttgart, Germany. The aim of the system is to provide a flexible and robust system for controlling a manufacturing line. This process has several basic parts where various operations are performed by different machines.

The industry-standard approach to manufacturing control is to devise a global schedule, typically covering one day, for the entire manufacturing process. That specifies when various parts should be released from their stores, which machines they should be routed through, and what operations should be performed by various machines. A problem with this centralized and pre-planned approach is that the planning is divorced from its execution. A schedule can rarely be exactly adhered to in practice as machines and operations fail in ways that are difficult to predict, and many operations take longer than expected. When such disturbances occur, plant manager must either initiate a rescheduling exercise or make an ad hoc decision. Both of these options lead to inefficiencies.

An agent-based approach was adopted to improve responsiveness, contingency handling, and fault tolerance. In this system, each manufactured part is represented by an autonomous agent that has the objective of getting itself to the end of the manufacturing line after a specified set of operations has been performed on it. Each machine is also represented by an agent. Such agents have the objective of maximizing their throughput, and they do that by deciding what parts will be accepted in what order and what operations will be performed at what time. Thus, for a given part to have an operation performed on it, its agent must negotiate with a machine agent capable of performing that operation. In short, resources are allocated dynamically on a just-in-time basis through a continuous coordination process among the relevant agents. The system has been running successfully with a performance rating of around 99.7 % of its theoretical optimum.

7.7.2 Designing of a Smart Grid

Several major power outages including the massive ones in 1996 and 2003 that affected parts of USA and Canada were caused by a chain of events that were set into motion by comparatively simple localized failures. As the power grids in North America are closely connected, it was found that they are quite vulnerable to such local failures. That led to the idea of increasing grid reliability and efficiency by embedding localized intelligence in a distributed power system.

The Main Characteristics of a Smart Grid

- Self-healing
- Attack resistant
- Consumer friendly
- Fulfill today's needs for electric power
- Able to accommodate all types of generation and storage options
- Optimizes assets and efficiency of operation

The new grid architecture will operate in a bottom-up manner, where distributed field equipment will tell the grid's higher management functions on how and where electrical currents are to be supplied. This is in contrast to the old-style top-down approach where decisions were imposed from above. That led to the use of independent localized intelligence, which can only be met by the use of agent technology. Research and development of systems using multiple agents for their use in smart grids is now in progress at a number of academic and industrial sites.

7.7.3 Buying and Selling of Electric Power

Electric power industries in many countries in the world are moving from regulated monopolies to competitive enterprises. With deregulation, customers are learning to look for the most cost-effective power available at a given time while producers and suppliers are trying to maximize their returns on investments. The industry is now adopting a transparent Internet-based electronic scheduling and trading system that provides information to all the market participants. In this environment, intelligent agents are becoming an integral part of the process, where they act as buyers and sellers of electric power in online auctions.

A research project conducted for the Electric Power Research Institute (EPRI) in Palo Alto, California demonstrated how intelligent agents could help producers, suppliers, and consumers of electricity to optimize their buying and selling process in this spot market. In this project, an agent may take on different roles at different times, such as being a buyer or a seller of electricity. An intelligent agent may advertise that it is going to conduct an auction and specify the time. Bidders interested in participating can then join and submit bids.

An agent may exhibit its unique and individual characteristic as determined by the needs of the party it represents, such as:

- Anxious buying and selling behavior, cool-headed behavior, or greedy behavior
- Operating, policies that dictate how much it is willing to spend or charge for power
- Ensure that adequate amounts of power are bought and sold to sustain operations

The research showed that distributed, multiple-agent approach is significantly more useful than the use of a central server, since individual agent's behavior is under the control of its stakeholders and is not subjected to the rules and constraints of a central authority.

Other successful examples of industrial intelligent agent-based applications include an automotive pattern shop in Czech Republic and packaging technology in Greece. Statoil of Norway is planning to use intelligent agent-based technology to help in optimization, planning, and process control in their trading and operation areas.

7.8 Agent-Based Modeling and Simulation

In conventional modeling, the cause and effect relationships are between classes or groups but not between individuals. Such a model deals with aggregates rather than with individual entities. Economists and social scientists often make the simplifying assumption that all individual objects within a class are identical or very similar in their actions and behavior.

Agent-based modeling and simulation is gaining popularity because it allows specification of individual entities and their interactions. An agent may be a person, or any other single object whose individual behavior pattern can be represented easily. There can be direct correspondence between agents and real world objects, which makes it easier to design a model and interpret the results of simulation. Agent-based models may also have memory and are able to simulate learning at both individual and group levels. A list of agent-based modeling packages along with details of NetLogo, a popular simulation package, is included in Appendix VI.

An example of agent-based modeling and simulation is included in Appendix VII. It shows that small levels of preference or restrictiveness can unwittingly lead to large levels of segregation between diverse groups. Whereas, a little amount of positive action can lead to a more balanced situation. A simulation of the example is available on the NetLogo website for readers to try out.

8

Discrete Events and Procedures

Abstract In previous chapters, the focus was on system behavior based on continuous interactions, where feedback loops predominate. However, it is quite common for a system to follow procedures that include ordered sets of tasks. Baking a cake, a monkey eating nuts, a mother conceiving and then giving birth to a baby follow certain sets of procedures. In industry, many processes such as manufacture of pharmaceuticals, fine chemicals, and beverages are largely based on sets of procedures. On the surface, procedures look quite simple, but they interact with continuous functions, thus, affecting the behavior of systems in various significant and interesting ways.

The chapter introduces procedural function with some simple examples. That is followed by discussions of such functions in social and natural systems and taking care of abnormal situations. Then there are discussions on the main characteristics of procedural functions. Finally, modeling and simulation of procedural functions are outlined.

8.1 Why Study Procedures?

Any discussion on systems is not complete without taking into account discrete events and procedures and how they affect their behaviors. Continuous functions work continuously and generally do not have definite start or end, whereas, discrete steps define actions that have beginning and end. An ordered set of discrete steps that define a required task or function is called a procedure. When I start my car, for example, I follow a set of discrete steps, such

A. Ghosh, *Dynamic Systems for Everyone*,
DOI 10.1007/978-3-319-43943-3_8

as opening the door, getting on the driver's seat, inserting the ignition key, depressing brake pedal, and then turning the key. An important point here is that the order of these steps does matter, for example, it is not possible for me to turn on the ignition key before opening the car door.

Similarly, when my daughter bakes my favorite cake, she follows a set of steps in a procedure, such as:

1. Set the oven at right temperature
2. Mix right amount of ingredients in proper order
3. Put the batter mix in the oven
4. Set the timer for baking
5. Wait for the timer to time out
6. Check if the cake is baked
7. If not done, then return to step 4 for baking for an additional period of time
8. Switch off the oven
9. Take the cake out
10. Allow the cake to cool before serving.

So far, the discussions have been limited to the behavior of systems based on continuous interactions of their subsystems and other systems around them, where feedback loops predominate. Whereas, in baking a cake the baker follows a set procedure, which to a large extent, does not depend on feedback signals during its execution. As stated before, unlike a set of continuous actions, a procedure has a definite start and an end along with defined steps and other logical requirements that must be followed in right order to perform the task correctly.

On the surface, a procedure looks quite simple. A main reason for discussing them here is because of their close relationship with continuous functions thus affecting the behavior of many systems in various significant ways.

Procedures are common in our day-to-day activities, such as buying a train ticket from an automatic dispenser, paying a bill, or sending an email to a friend. They are also common in engineered, political, economic, and social systems. Birds, insects, and other animals follow procedures for gathering food, for mating, and for home and nest building.

Capuchin monkeys in Brazil, when hungry, follow a procedure to crack palm nutshells to eat the meat. To do so, a monkey climbs a palm tree and taps a nutshell to make sure that it has enough nuts in it. It then picks one that has enough nuts and throws it to the ground. After climbing down, the monkey nibbles away the soft outer covering of the nut until the woody shell is exposed. Then the monkey attempts to crack the shell by striking it with a piece of rock. If successful, it eats those nuts but if the shell is too hard to crack, it leaves it on the ground for a few days to allow the shell to rot partially, making it easier to crack (Figs. 8.1 and 8.2). (See Appendix VIII for details on the documentation of procedural functions).

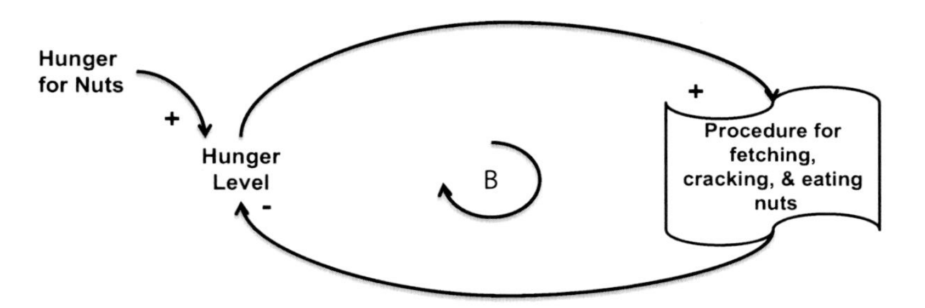

Fig. 8.1 A causal loop diagram with a procedure

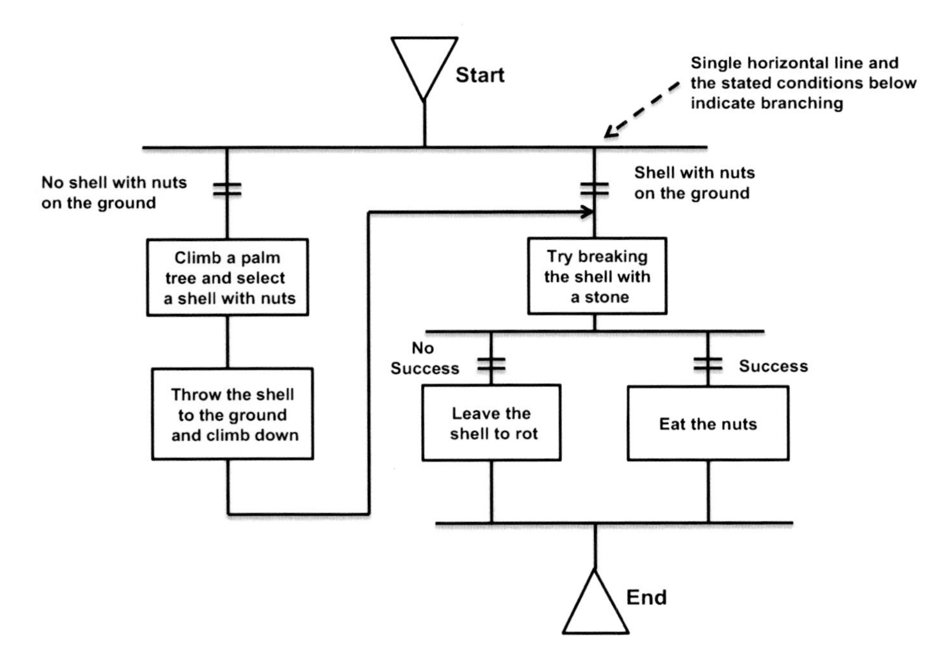

Fig. 8.2 A procedure function chart for fetching, cracking, and eating nuts

8.2 Procedures in Engineered Systems

In Chap. 1, we briefly discussed a home-heating system, where a feedback loop helps keeping a room warm at a constant temperature. If we now add an air conditioner to keep the room cool, during summer months, then we need a set of logical steps to keep either the heating or the cooling system active, depending on the weather conditions. A single control loop for controlling both heating and cooling systems will not work well as it may alternately activate the two systems during the change of season, causing the room temperature to oscillate. In order to avoid that situation, we may define a set of logic for activating either heating or cooling system based on outside temperature (Fig. 8.3):

- If outside temperature is greater than 75 °F, then deactivate heating system, activate cooling system, and set the thermostat at 72 °F.
- If outside temperature is less than 60 °F, then deactivate cooling system, activate heating system, and set the thermostat at 68 °F.
- If outside temperature is between 60 and 75 °F, then deactivate both heating and cooling systems.

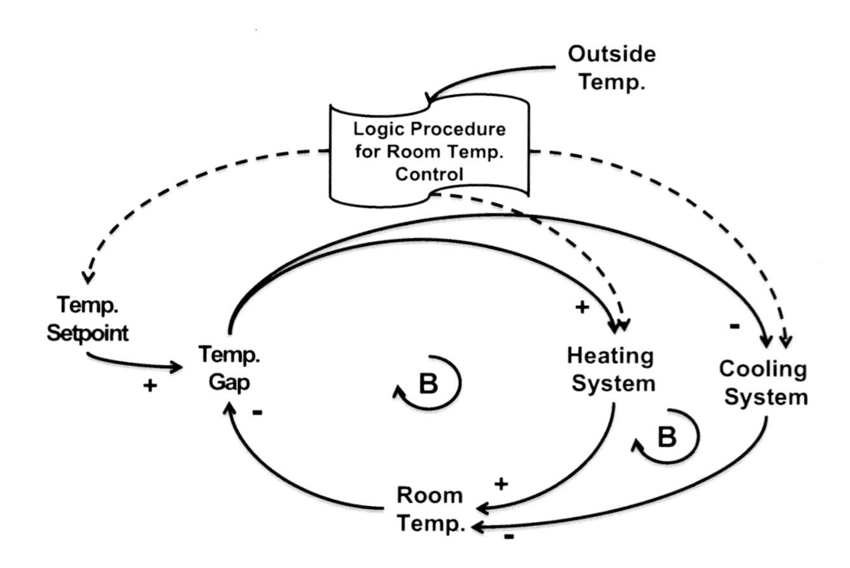

Fig. 8.3 Room temperature control with logic procedure

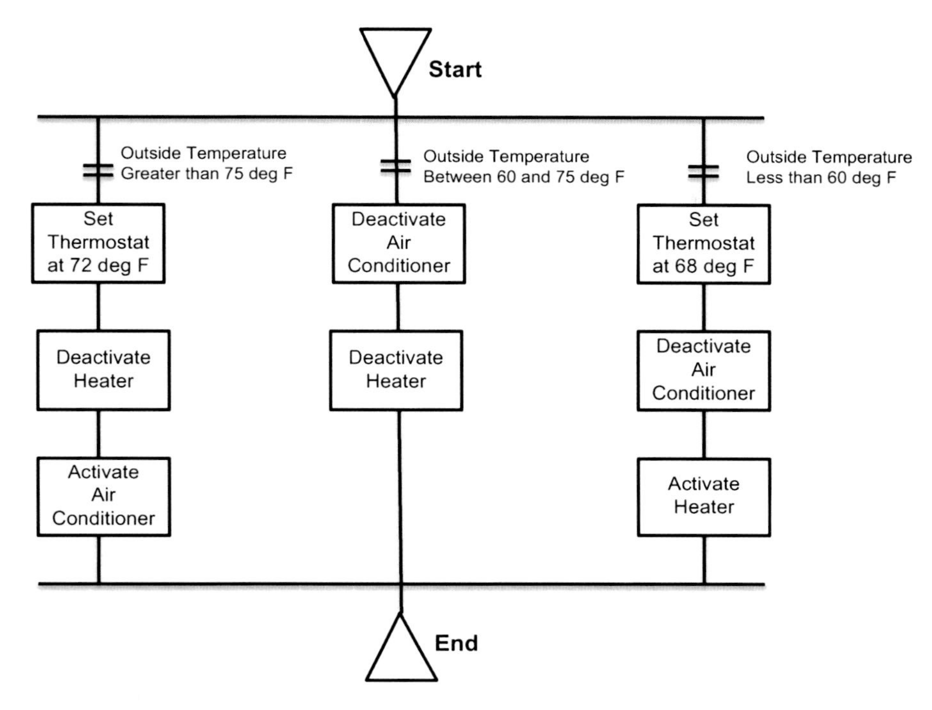

Fig. 8.4 Logic procedure for a room temperature control

This simple logic may be depicted by a procedure function chart (PFC), where the logic specifies conditional branching depending on the outside temperature on a given day (Fig. 8.4). This logic may be manually started, when needed or may be made to run periodically.

In many engineered systems that are in essence continuous, procedures are often used for their startups and shutdowns, and for taking appropriate actions when they malfunction. Thus, a distillation unit in an oil refinery runs mostly on feedback control during its normal operation, but requires one or more discrete procedures for its startup, shutdown, and for addressing exception or failure conditions. There are, however, other industrial processes, where procedural control predominates, such as in manufacturing drugs, fine chemicals, paints, food, and beverages, where products are produced in batches. These processes are essentially discontinuous and are commonly called “batch processes.” However, there are many continuous functions with feedback loops in these batch processes. Maintaining proper temperature and pressure for a period of time, for example, is accomplished by employing feedback control. Thus, procedural and continual functions are deeply intertwined in many systems.

In Appendix IX, there is a detailed discussion on a chemical-batch-manufacturing process where procedural functions predominate. By studying the underlying principles of such a process, one is in a position to apply similar methods for analyzing and designing procedural functions in other areas, such as social, legal, and political systems.

8.2.1 Formula Variables

Formula is a set of data that a procedure may need to make a batch of product. The formula for a cake, for example, may include quantities of flour, butter, sugar, milk, and eggs that are needed to make the batter. It may also include baking temperature and time. That makes the procedure more flexible, where a procedure may be used to make different types of cake by selecting appropriate formula.

In the example of room temperature control, outside temperature limits and set points are formula variables. Thus, formulas are extensively used in industry for manufacturing different but similar products and for different grades of same product. In natural world, each living being has a set of DNA, which like a set of formula variables specifies many of its characteristics.

8.3 Procedural Functions in Social Systems

Preceding example illustrated the use of procedure for a home-heating system; however, similar concepts may be applied in numerous daily activities. Grocery shopping, for example, is a procedure consisting of a number of sequential steps, such as:

1. Make a grocery list
2. Drive to a grocery shop
3. Find the listed items and put them on a grocery cart
4. Pay at the check out counter
5. Transfer the items bought from the cart to the car
6. Drive back home
7. Unload the grocery
8. Store the items in proper bins and shelves.

Each of these steps may consist of a series of steps at a lower level. Feedback or causal loops do not play a big role in the execution of these sequential steps at this level, but that may change as one gets into lower levels of action, such as driving the car. Then the shopper needs to keep a continuous watch on traffic conditions.

Procedures are used extensively in political, economic, and social systems. In the USA, for example, when arresting a suspected felon, a police officer must follow the law by reading Miranda rights at the proper time. If that step is skipped or not done correctly, the person arrested could be freed on a technicality. A procedure needs to be followed for obtaining a motor vehicle drivers license. That includes paying a fee, taking an examination on an interactive computer terminal, passing a vision examination, and passing a road test. In every country, there are extensive sets of laws and accepted sets of procedures that govern the activities of individuals, groups, and institutions. Even people living in primitive societies, with no written language, need to follow many customs and procedures that are handed down from one generation to the next.

8.3.1 Taking Care of Abnormal Conditions

In the example of grocery shopping, the shopper has established a procedure, which the shopper follows every weekend. That includes making a list of groceries, driving to her favorite store, do the shopping, and then driving back home. While she may try to follow the same procedure every time she goes for grocery shopping, what happens when one of the tires of her car goes flat while driving to the grocery store? In that occasion, a set of exception logic kicks in, she stops the car at the roadside to replace the wheel with flat tire with the spare, which takes a considerable time and effort. She then has to decide, either to continue with her grocery shopping procedure or abandon it to get back home and start it all over again later.

Similar is the case with an industrial batch process where, if something goes wrong, exception logic is initiated to take care of the situation. Once the situation is under control, depending on the severity of the problem, the normal procedure is either resumed or terminated by draining out the unfinished product.

8.4 Procedural Functions in Natural Systems

It seems that most of the natural systems operate on feedback, which may be depicted by causal loop diagrams. However, on closer examination, we do find processes that operate like sequential procedures.

Examples of Procedures in Natural Systems

- Deciduous trees lose leaves at the onset of winter
- Birds migrate with the change of seasons
- Cells in our bodies age with time
- The inherited genes specify the characteristics of an offspring, where the DNAs act like formula variables
- The circadian rhythm in our body that drives the 24-h cycle, which requires resetting when traveling across time zones.

When we are growing, the cells in our body multiply. They also regenerate to keep us fit, but the capacity for regeneration declines with age. The progression of life from birth to growth and then adulthood, old age, and death follow a sequence, which is very much akin to a procedure. Similar is the case with biological inheritance, where genetic codes passed on by parents specify the characteristics of an offspring. These processes, like ageing of our cells, reproduction process, and our daily body rhythm, are like sequential operations, though there is room for further research in these areas.

8.5 Main Characteristics of Procedural Functions

A Procedure Specifies a Specific Sequence of Actions for Accomplishing a Task A procedure is a step-by-step plan of actions for the completion of a task. It may be a written document, a graphical or pictorial diagram, a table, or just a mental idea. If a procedure is properly executed, the intended task should be complete.

A Procedure Consists of a Specific Sequence of Steps Each step in a procedure specifies an action, which for a simple situation seems straightforward. However, for most practical situations, where there are many actions with decision points and actions running in parallel, it can get quite complex.

A Procedure Has a Beginning and an End with Loopbacks, Where Needed A procedure specifies a way to complete a task. A procedure, unlike a continuous function, has a beginning and an end. However, a procedure may loop-back to the beginning or to an intermediate point, when a set of tasks need to be repeated.

A Set of Procedures May be Configured as a Hierarchy A set of procedures may be made up of one procedure at the top and a number of lower-level procedures. For a long or complex task, one or more levels of subprocedure may be specified. This concept of a hierarchy is quite useful, as it reduces complexity in specification and execution of a task.

In an Automated System, Only the Procedure at the Highest-Level Needs to be Initiated Manually Generally, procedures do not initiate themselves, they are either initiated manually or by another procedure, or by a system or subsystem. In a hierarchical procedural system, lower-level procedures are generally initiated by higher-level procedures. Exception to this rule may be found in an intelligent-agent-based system where procedures are self-initiated based on external conditions.

Abnormal Conditions May Put a Normal Procedure on Hold and Initiate Exception Procedure(s) Normal procedures for completion of a task may be interrupted by unforeseen events. In baking a cake, for example, electric power that heats the oven may fail, requiring corrective action like extending bake time when the power is restored or throwing away the batch if the power failure is for an extended period. In an automated system, exception procedures are often specified for known failure conditions, but human judgment is needed for unforeseen situations.

A Procedure is Generally Static During Its Execution A procedure is generally not changed or altered during its execution. A procedure, by its nature, is something that has been tried and proven, thus, requiring infrequent changes and updates. However, a procedure is not sacrosanct; it may be altered and updated to keep pace with societal, environmental, and other changes. Generally, there is a limited room for updating or modifying a procedure while it is running. Significant revisions and optimizations are carried out only after the completion of a run.

Same Procedure May be Used for Similar, but Different Systems The advantage of using an established procedure is that it can be used repeatedly to accomplish similar tasks. The procedure for cracking nuts, for example, is repeated not by a single monkey but also by all members of the tribe. In a batch process, a procedure may be used repeatedly for manufacturing similar products. The constitution and the legal system of a country are the basis of many procedures that need to be followed by every citizen.

A Detailed and Accurate Documentation of a Procedure Is an Arduous Task Mental models of procedures are comparatively easy to make and maintain, but they are not always accurate. Generating a comprehensive and accurate procedural document can be a quite daunting. In some situations, such as in production of drugs and food products, there are legal requirements for accurate documentation of procedures for manufacturing a product. Various languages are used for documentation, which include—sequential function chart (SFC), procedure function chart (PFC), flow chart, and Structured English (Appendix VIII).

A Procedure May Set Initial Values or Change Current Values in a System In a batch process, a formula table specifies a set of values that are needed for the production of a batch. These variables may be batch size, set points, time for cooking or reaction, and the like. However, procedures may also be used for changing the grade of a product in a continuous process, such as changing the thickness or quality of product in a paper machine. In a crude oil distillation unit, the proportion of various distillates (e.g., diesel fuel and gasoline) may be altered to meet market demands by changing set points of a number of control loops. These changes are accomplished in an ordered manner by prewritten procedures.

8.6 Importance of Discrete and Procedural Functions

In the past, continuous and procedural functions were often treated as two very different types of behavior. They were designed, analyzed, and studied as two distinct disciplines. However, an in-depth examination shows their close relations. While analyzing or designing a system one should not only focus on its continuous functions but should also include its discrete and procedural parts. That will enhance the overall understanding of a system's behavior. Hence, procedural function should be given its due importance by including it as a part of normal curriculum of a system dynamics course.

8.7 Modeling and Simulation of Procedural Functions

Appendix VIII describes some common methods for specifying and documenting procedural functions. Simulation packages such as iThink/Stella, Insight Maker, and NetLogo are well suited for simulating, continuous functions but they are not designed for simulating procedures. MATLAB and Simulink software packages allow discrete event simulation with a limited amount of procedural functions.

Major control-system suppliers have been offering software packages for control and simulation of both continuous and procedural functions for a number of years, but they are not well suited for general use. A software package for general use called AnyLogic, developed by XJ Technologies in St Petersburg, Russia, allows users to simulate continuous, discrete/procedural, and agent-based simulation on a single platform. The package was first released in the year 2000 and has since then gone through a number of upgrades.

9

Unintended Consequences

Abstract Any action to manipulate a system to produce a desired outcome can generate one or more unintended consequences. We all know that a number of powerful and effective drugs for reducing or curing disease also produce harmful side effects. While scarcity and high cost of food has led to malnutrition and starvation in many underdeveloped economies, solving of food problem is leading to obesity and other related ailments. With proper application of system knowledge, we can reduce or eliminate such negative consequences. However, not all unintended consequences are bad, as is the case with aspirin, which was originally used for pain relief but was later found to be an anticoagulant that reduces the possibility of heart attack.

The chapter offers many examples of unintended negative and positive outcomes of our actions. They range from cobra effect to iron ore mining boom in Australia. The chapter ends with a discussion on ways to minimize or eliminate these unintended consequences.

9.1 Unintended Negatives and Positives

Unintended consequences are unexpected results of our actions, which are often detrimental. Sometimes they are called side effects, as in the case of a drug, which is effective in treating an ailment, but may also produce undesirable symptoms in a patient. In a complex system with many feedback loops, one cannot always be sure of all the outcomes of a deliberate action. Robert K. Merton, a well-known sociologist in a paper published in 1936,

A. Ghosh, *Dynamic Systems for Everyone*,
DOI 10.1007/978-3-319-43943-3_9

"The Unanticipated Consequences of Purposeful Social Action" laid groundwork on this topic, where he listed five main reasons:

- Ignorance—impossibility of anticipating everything, leading to incomplete analysis
- Error—incorrect analysis
- Immediate interest—addressing immediate symptoms rather than long-term causes
- Basic values—basic beliefs and values of an individual or a society, which inhibit taking measures that are appropriate
- Self-defeating prophecy—acting on a problem that may occur in the future leading to an increased possibility of that problem's occurrence.

In 1971, Professor Jay W. Forrester published a paper entitled "Counterintuitive Behavior of Social Systems," where he stated that human mind is not adapted to interpret readily the behavior of social structures, which are multi-loop nonlinear feedback systems. However, he stated that proper development and use of social system models could lead to a better understanding of their structures and their behaviors.

9.1.1 Unintended Negatives

There are many examples of unintended negative consequences. For example, rent control for properties in many major cities is intended to make housing more accessible for lower income tenants. However, that can also reduce the quantity and quality of housing. Property owners are often reluctant to maintain or improve their properties that are under such control because there is little financial incentive to do so. In some places, property owners circumvent rent control by extracting cash bribes, which is termed as key money.

Rapid technological development in consumer goods, such as computers, digital cameras, and telephones, is offering many beneficial effects like increased choice and affordability. However, that often leads to higher levels of anxiety and dissatisfaction for the consumers who find it more difficult to choose the best items that make economic sense and fulfills their needs. Additionally, a customer may find that the best price he or she pays today is no longer true tomorrow. Thus, rapid technological advance and increased choice in consumer goods, which are positive developments, are also leading to negative consequences. Looking around, we find many more examples of unintended consequence, such as low tar cigarettes leading to higher consumption by smokers, pesticides producing resistant pests, and large-scale use of antibiotics producing resistant strains of bacteria.

9.1.2 Unintended Benefits

While a majority of unintended consequences have negative impacts there are some that turn out to be positive. Aspirin, for example, was originally used as a pain reliever, but was later found to be an anticoagulant that helps in preventing heart attack, reduces severity and damage from thrombotic stroke, along with other benign properties.

Today pollution, disease, and overfishing are causing serious negative impacts on coral reefs. While we are pushing many coral reefs around the world to the point of destruction, but by design or accident are also creating new ones. Ships sunk in shallow waters during wartime have created the base for many coral reefs, which are scientifically valuable and are proving attractive to recreational divers.

9.2 Examples of Unintended Consequences

Here are some examples of unintended consequences of human actions or natural catastrophes. They illustrate that not all of them have negative outcomes.

9.2.1 The Cobra Effect

The term "Cobra Effect" comes from an anecdotal story set at the time of British rule in India. When authorities got concerned about the number of venomous snakes in Delhi, it offered bounty for every dead cobra. The strategy was initially successful as large numbers of snakes were killed for rewards. Soon, however, breeding cobra became a profitable enterprise. After some time the reward program was scrapped, as the authorities came to know about the breeding plans of cobra catchers. That caused the breeders to set free the now-worthless cobras, which led to a net increase in cobra population. Thus, apparent solution for a problem made the situation much worse than before.

A similar event occurred in Hanoi, Vietnam, under French rule. A program was created for getting rid of rats in the city. The program paid a reward for to anyone who killed a rat and produced its severed tail as evidence. However, officials soon began noticing rats in Hanoi without tails. The rat catchers were capturing rats, cutting off their tails, and then releasing them back so that they could breed more rats, thus augmenting the possibilities of making more money.

Noted German economist Horst Siebert wrote a book in 2001 titled "The Cobra Effect" to illustrate the causes of incorrect stimulation in economy and politics. It is now used as a common term to caution people against using reward to solve a problem. It also draws attention to the fact that apparent solutions for many common problems do not work because of their unintended consequences.

9.2.2 Prohibition in the USA

At the start of prohibition era in the USA, it was expected that the money that went on to purchase booze would be diverted to clothing and other consumer goods. Property owners expected rents to rise as saloons closed and neighborhoods improved. The real estate developers expected property values to rise. Entertainment industries, like theater producers, expected more clientele as people would look for new ways to amuse themselves without alcohol.

However, the unintended consequence was a general decline in amusement and entertainment industries. Many restaurants failed because they were no longer profitable without liquor sales. Theater revenues declined rather than increased, and few of the other economic benefits that had been predicted did not happen. Closing of breweries, distilleries, taverns, and bars led to reduction or elimination of many occupations, such as barrel making and trucking. Thus, the initial economic effects of prohibition were largely negative.

One of the major consequences of prohibition was its effect on government tax revenues. Many states, such as New York, relied heavily on excise taxes on liquor sales to fund their budgets. It cost the federal government in the tune of $11 billion in lost revenues, while costing over $300 million to enforce the prohibition. That led to a significant increase in income tax rates at federal and state levels to balance their budgets.

While the Eighteenth Amendment of US Constitution prohibited manufacture, sale, and transportation of intoxicating beverages, it did not outlaw possession or consumption of alcohol in the USA. That and the way of prohibition were enforced led to many other unintended consequences. For example, the law allowed pharmacists to dispense whiskey by prescription for a number of ailments, ranging from anxiety to influenza. That led to a dramatic increase in the number of registered pharmacists in New York State as bootleggers soon found that running a pharmacy was a perfect front for their trade. Enrollments to churches and synagogues rose as public were allowed to take wine for religious purposes. There was also a large increase in the number of self-professed rabbis, who could obtain wine for their congregations.

The law was unclear as to whether Americans can make wine at home. That led American grape industry to sell kits of juice concentrate with instructions for making wine. Additionally, the dubious quality of alcohol on black market became a serious health hazard.

Effects of prohibition also had a very negative effect on police officers and prohibition agents as many of them were tempted by bribes and some of them found it lucrative enough to start bootlegging themselves. While many stayed honest, there were enough of them who succumbed to the temptation thus undermined public trust in law enforcement.

Growth of illegal liquor trade made criminals of millions of Americans and the courtrooms and jails overflowed. As the backlog of cases increased, many defendants had to wait for a long time to be brought to trial. However, the greatest unintended consequence of prohibition was that the law that was meant to foster temperance led to intemperance and excess.

9.2.3 War on Drugs

Many countries such as Mexico, Columbia, and the USA are both producers and consumers of illegal drugs. They are currently spending enormous amount of money and resources on a war on drugs, and the cost is escalating.

The war on drugs has not achieved its intended objectives, while it has triggered a long trail of unintended consequences. These are possibly causing more damage to the societies than the drugs themselves. The cost of prohibition seems to be borne disproportionately by developing countries that traditionally grow the crops that are used for the production of drugs, or they serve as trade routes to drug consumers in rich countries.

While the war on drugs has reduced cultivation of illegal drugs in some areas, it has also led to incarceration of a large number of drug users and intermediaries at a great cost to the governments. The number of adults incarcerated for drug law violations in the USA increased by more than ten-fold between 1982 and 2002, with the result that nearly one-fourth of the prison population today is for drug-related offenses.

The unintended consequence that stands out is the steep escalation of street price of drugs leading to increased profits for producers and, more importantly, to intermediaries who sell them (Fig. 9.1). Large profit margins have given rise to criminal organizations (cartels) that use violence, insurgency, and corruption to stay in business. Emergence of rich and powerful cartels has led

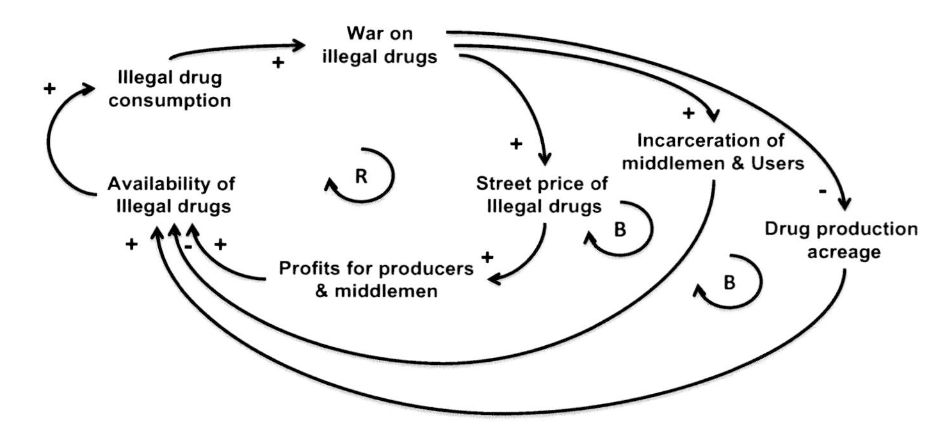

Fig. 9.1 The war on illegal drugs

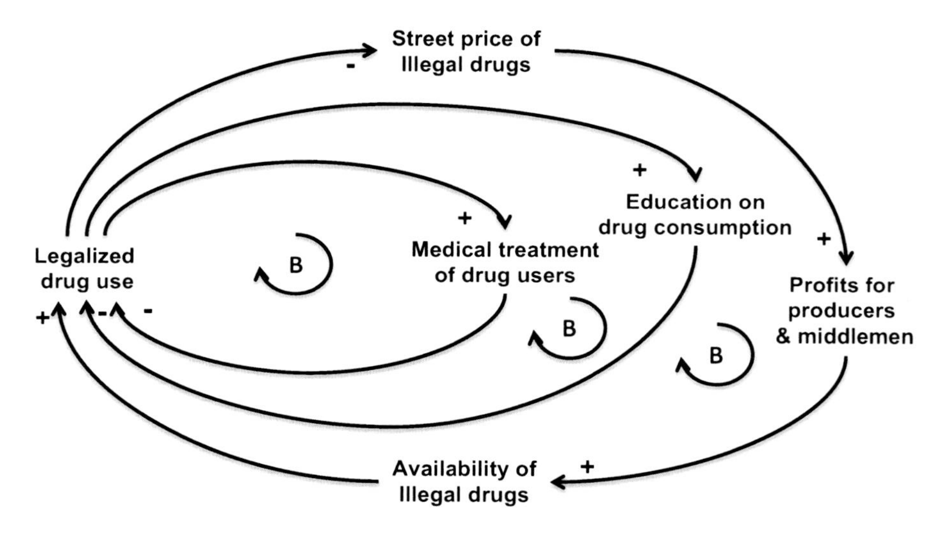

Fig. 9.2 Legalized drug use

to increased corruption of police and customs officials. In addition, large-scale corruption of some politicians and justices is undermining the stability of some drug-producing countries.

Thus, there is now a strong argument for changing drug policies, where education and medical treatment need to be emphasized rather than outright prohibition (Fig. 9.2), which include:

- Aggressive educational campaigns to make drug use less socially acceptable
- Changing the status of addicts from drug buyers to patients, to be treated by public-health care systems

- Controlling the quality and quantity of drug consumption by the addicts by dispensing them as prescribed medications.

These actions will reduce the profitability of drug cartels, leading to their decline and possible demise. These policies, however, need to be fine-tuned for each country in a way that is consistent with its history and its prevailing culture.

9.2.4 Mysterious Killer of India's Vultures

Vultures play an important role in the disposal of dead animals in the Indian subcontinent and around the world. Vultures are also an integral part of the Parsi (Zoroastrian) custom of disposal of dead bodies, where human corpses are left out to be consumed by them. Only a decade back one could observe numerous vultures roving around on the skies in the Indian subcontinent but suddenly they started to get scarce. Hide collectors usually remove the hide of dead animals leaving the vultures to take care of the rest. However, with their decrease in number, stray dogs and other wild animals have taken their place, leading to an increase in rabies and other communicable diseases.

The cause of disappearance of vultures was a mystery for a number of years until the completion of an intensive study of vulture colonies. The researchers found that a large number of vultures are dying because of kidney failure. After further investigations, they found that the root cause of this problem was the buildup of a drug called diclofenac in their tissues. Diclofenac is a generic nonsteroidal anti-inflammatory drug (NSAID) also known as Voltaren, which is usually taken by human beings to reduce inflammation and pain due to arthritis and other similar conditions. In South Asia, where a number of animal species, such as horse, ox, and buffalos, are used for tilling the soil, carrying loads, and drawing carriages, diclofenac was found to be a cheap and effective remedy to treat their sore or bruised limbs. The unintended consequence of the use of diclofenac for domestic animals is that the vultures are getting fatal doses of this drug when consuming their carcass.

These findings led to the ban of diclofenac for veterinary use in India, but that is being circumvented by many farmers and other villagers, as the drug is still readily available for treating human ailments. This shows that a substance that has benign effects on one species can produce very negative consequence on another.

9.2.5 The Great Fire of London

The great fire of London started in a baker's shop on September 2, 1666 and lasted for about 4 days. It destroyed over 13 thousand houses and a large number of churches including the original St. Paul's Cathedral. Londoners, at that time, were not conscious of the dangers of fire outbreaks. There was no organized fire brigade and the houses were largely built of timber and covered with pitch or asphalt. The houses were also built close together with little space between them allowing the flame to spread easily from one to another.

The fire was preceded by a long dry summer that caused draught in the city leading to scarcity of water to fight the fire. Lacking organized fire brigades, people tried to put out the fire using leather buckets but with little effect. The Mayor of London ordered that the houses near those on fire be pulled down but that made little difference. Finally, the Royal Navy used gunpowder to destroy a number of houses that were in the fire's path, thus, successfully stopping its spread. The great fire spread over more than 400 acres destroying about four-fifth of the city.

What is remarkable is that the fire claimed only 16 lives. But what is more noteworthy is that it saved many more lives by stopping the spread of black plague, which started a year earlier, as the fire killed most of the rats that were spreading the disease. The devastating fire also led to the reconstruction of a major part of London, which was directed by Sir Christopher Wren, the great architect at that time. He oversaw the reconstruction of many churches that were destroyed including St. Paul's Cathedral. Bricks and mortar instead of timber were mostly used for the construction of new houses. Additionally, congestion was reduced by widening and straightening some of the roads. The fire created an opportunity to rebuild the central area of the city, which subsequently became the hub of British Empire. Thus the fire, which caused a lot of financial damage and hardships to residents of London, showed that sometimes a highly negative incident could also lead to a number of positive outcomes.

9.2.6 Viagra Leads to Wild Life Conservation

In many traditional societies, body parts of many animals such as tigers, seals, and bears are considered aphrodisiac. These organs fetch huge price in markets leading to slaughter and possible extinction of some of the species. The good news is that the introduction of Viagra and other similar drugs are reducing demand for these traditional potions. These synthetic drugs are much cheaper, easier to obtain, and are more effective than traditional remedies with dubious efficacy.

It is estimated that demand for animal-based aphrodisiacs has fallen by over 70 % in many regions, which is a very good news for animal conservationists. Wildlife conservation agencies may consider offering free supplies of Viagra or similar drugs to people in regions where traditional aphrodisiacs are still in demand. Though many scientific discoveries have unintended consequences that may be disastrous, the discovery of drugs to address erectile dysfunctions is leading to positive consequence in the conservation of wild lives.

9.2.7 Negative Consequences of Increased Sanitization and Sterilization

We normally associate dirt and germs with various forms of sickness and disease. That includes communicable diseases like common cold and influenza, mosquito-borne diseases like malaria and dengue, waterborne diseases like cholera, and autoimmune diseases like asthma, hay fever, and allergies.

We can trace this thinking back to John Snow who postulated "germ theory," which proposed that microorganisms are the cause of many diseases. He used this to trace the 1854 cholera outbreak in London to germs in drinking water. The germ theory was later confirmed by Louis Pasteur and Robert Koch. Since then, there has been a push to get rid of germs from our environments, from food we eat, air we breathe, and articles we come in contact.

Several public-health measures have been taken by developed countries to reduce germs in human environment. This has led to decontamination of water supply, sterilization of milk and other food products, vaccination against common childhood infections, and wide use of antibiotics. Public awareness of the connection between germs and sickness has led to a huge demand for antiseptic toiletries, hand sanitizers, floor cleaners, and various other cleaning agents for everyday use. The increased hygienic awareness among individuals and societies has led to a decline in various communicable diseases in developed and developing parts of the world.

However, with decline of communicable disease, we are now experiencing rise of allergic diseases like hay fever, eczema, hives, and asthma. While only 10 % of the population in developed countries suffered from allergies two decades ago, today that has increased almost threefold. According to the American Academy of Allergy, Asthma and Immunology, about 50 million Americans now suffer from allergic conditions and the number is increasing (Fig. 9.3).

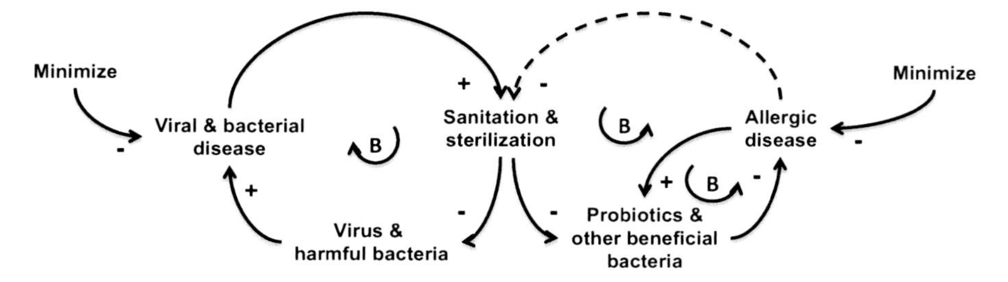

Options:

1 – Reduce sanitation & sterilization (not acceptable)
2 – Do selective sanitization & sterilization without affecting the probiotics (not possible)
3 – Produce and increase the intake of probiotics & other beneficial bacteria

Dotted arrow indicates options 1 and 2, which are not feasible at the present time.

Fig. 9.3 Negative consequences of excessive cleanliness

In countries where good health standard does not exist, people may be chronically infected by various pathogens, but the prevalence of allergic disease remains low. This shows an inverse relationship between the level of hygiene and the incidence of allergies and autoimmune diseases. The more sterile the environment a child lives in, higher is the risk that the child will develop allergies or an immune problem in his or her later life.

It is now strongly believed that the bacteria in our digestive system, that are essential to digestion also, serve to educate our immune system. They teach the immune system how to react to strange substances. Cleanliness does reduce our exposure to harmful bacteria but it also limits our exposure to beneficial microorganisms. As a result, the bacterial flora of our digestive system is not as rich or diverse as it used to be.

So, what is the recourse to this problem? Medical and public health community have been pondering on this issue for some time. They cannot recommend going back to lower hygienic standards of earlier days, which would lead to increased incidence of parasitic disease. Another course may be to kill harmful virus and bacteria while keeping benign bacteria intact. This requires performing selective sterilization, which is not possible with our present level of knowledge in this area. The third course of action is to increase the production of probiotics and other beneficial organisms and encouraging their intake, especially for babies and small children. A common source of probiotics is yogurt with live cultures. Some studies have found that women who ate yogurt in the last third of their pregnancy may significantly reduce the impact of allergies during the first two years of their child's life. Though taking probiotics may not be a complete solution to this problem, it seems to be a step in the right direction.

9.2.8 Aswan High Dam

After 11 years of construction, President Anwar Sadat of Egypt officially inaugurated the Aswan High Dam in January 1971. It was built to regulate annual flooding of Nile valley and to create a reservoir for storing water to prevent water scarcity during years of severe drought. Construction of the dam began in 1960 as a national project by President Nasser who died a year before its completion. The dam is located just south of the city of Aswan in Egypt, about 500 miles south of Cairo. The dam is about twelve thousand feet long, over three thousand feet wide and over 360 ft high. It has created a huge artificial lake, called Lake Nasser, which is around 300 miles long and 10 miles wide.

The project has brought many benefits to Egypt, including flood and drought control, a significant increase in agricultural land, and a huge supply of hydroelectric power. However, from its very inception, the project has been quite controversial. First, there was political controversy of cold war at that time and then there were known and unknown consequences of a project of this magnitude.

The political controversy had its origin in the USA's refusal to fund the project, which it had initially agreed to do. Soon after the withdrawal of American help, President Nasser decided to nationalize the Suez Canal Company, not only to provide funds for this project but also to restore Egypt's past glories and to safeguard national dignity and pride. Then, Soviet Union offered a low interest loan and technical expertise, which were readily accepted by Egypt. Thus, cold war politics produced significant media and public interest around the world, which contributed to heightened global awareness about this project.

No environmental impact assessment was made before or during the design and construction of this dam, as that was not required and its methodologies were relatively unknown at that time. However, a number of reports that were published then were very critical of its outcome. These media articles focused on negative environmental impacts of the dam, which found a receptive audience in Western countries. Many people were convinced that such a large development project could only bring environmental disaster. However, assessments that are more recent indicate that the project has been largely successful in uplifting the economy, without which Egypt would have been in a much poorer shape. It nevertheless had some adverse impacts (Fig. 9.4).

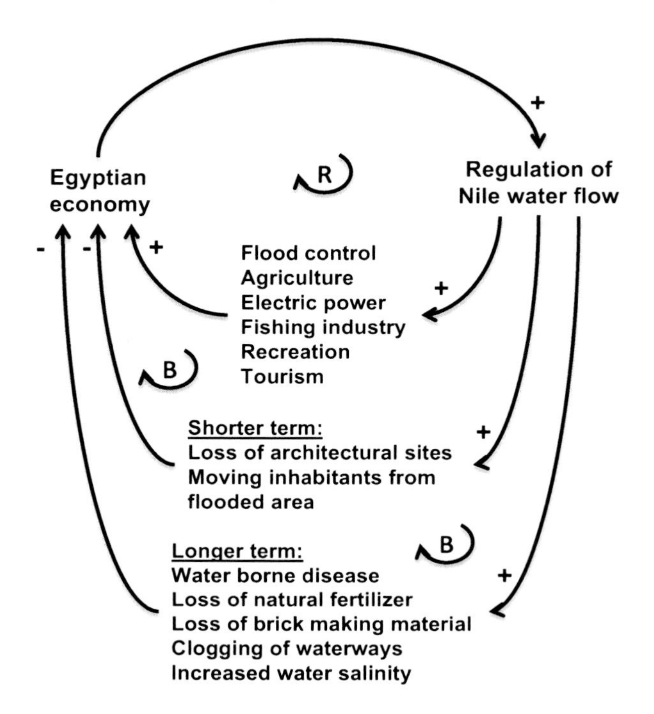

Fig. 9.4 Effects of Aswan High Dam on Egyptian economy

Positive impacts of the project:

- Provided a constant water supply to farmers
- Significantly helped in controlling floods
- Allowed farmers to grow multiple crops in a year, instead of just one
- Produced a considerable amount of electric power for industrial and domestic use
- Significantly increased the amount of available arable land
- Created a large freshwater lake, resulting in a thriving fishing industry and a great recreation area in an inland region
- Increased tourism along Nile River and throughout Egypt.

Detrimental effects that were addressed during the course of the project:

- As many archeological sites were flooded, some were moved to higher grounds at a great cost
- Many inhabitants in the area had to be relocated.

Detrimental effects that were not anticipated or ignored during the course of the project:

- Pressure of dammed water caused seepage to surrounding dry land, which increased the potential for earthquakes in those areas

- A significant amount of water loss due to evaporation
- Increase in waterborne diseases in humans and animals in downstream areas
- Increased demand for artificial fertilizers as water downstream carried less silt and nutrients
- Silt that was used to make bricks for building constructions is no longer available leading to excavation of agricultural land, thus, reducing crop-growing areas
- Greater growth of algae and other aquatic plants that clog downstream waterways
- Increased salinity of downstream water, especially near the delta regions, making it unfit for irrigation and drinking
- Possibility of dam failure, which may lead to major flooding and human catastrophe.

Despite these problems and concerns, most people feel that the dam has been a great success and has significantly helped Egyptian economy. The dam's economic benefits have never been in doubt. Increase in agricultural production, hydropower generation, and improved navigation along with savings from flood damages has paid off the cost of dam construction in 2 years. The real question is not whether Aswan Dam should have been built, but what steps should have been taken to maximize positive socio-environmental benefits while minimizing its negative consequences.

Aswan High Dam offers lessons to future dam builders on making stringent environmental impact assessments, implementing better policies for resettlement of displaced people, and applying new technologies to reduce other negative consequences such as reduction in natural nutrients from silts, increase in water borne disease, and clogging of waterways with the growth of algae and other aquatic plants.

9.2.9 Australian Iron Ore Mining Boom

Australia has a vast reserve of iron ore, which, until recently, was largely untapped except for a few local iron and steel mills. That has changed with China's huge demand for iron for infrastructure development and for its export-driven economy. The resulting rapid development of Australian iron ore mining has greatly increased revenues and profits of mine owners, which had been moribund for a long time. Along with that have come increased employment opportunities in this industry, leading to the lowest unemployment rate for any country in the developed world. Increased prosperity of

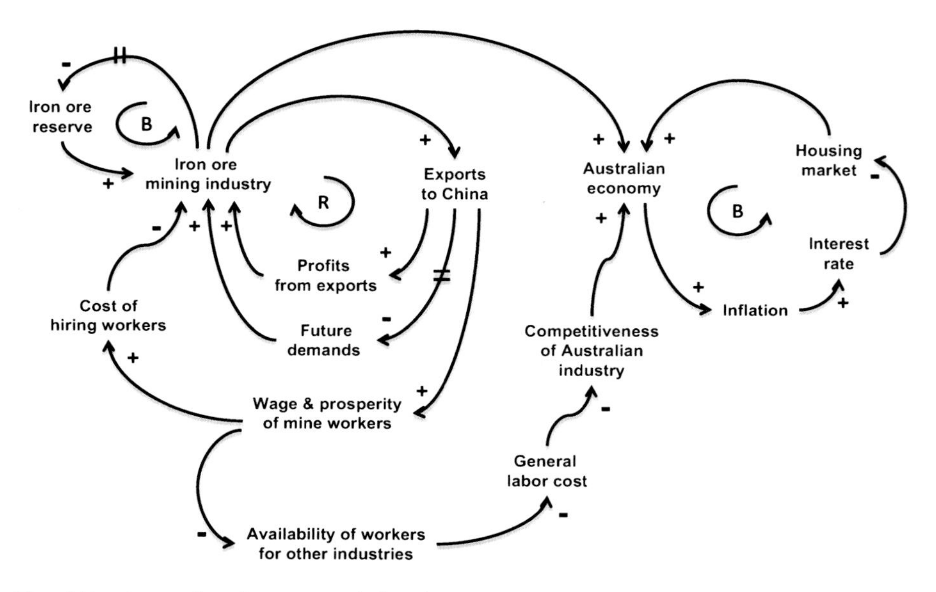

Fig. 9.5 Australian iron ore mining boom

mine owners (getting lion's share) and mineworkers has also led to a general improvement in Australian economy. That has resulted in the appreciation of its currency making imported goods much cheaper and affordable for consumers (Fig. 9.5).

The iron ore mining boom, however, has a number of dark sides that are largely being overlooked in the present state of exuberance. Some of the negative consequences are immediate, while there are others that are going to show up in the years to come.

9.2.9.1 Immediate Negative Consequences

One of the immediate effects of the boom is increase inflation rate, which has led Australian Reserve Bank to raise interest rate a number of times in past few years. Increased interest rate has led to a surge in mortgage cost resulting in a decline in housing market. Though the Australian currency appreciation has made imported goods cheaper it has made its manufactured exports that much dearer putting them at a competitive disadvantage.

Australian steel industry is no longer able to compete with imports leading to the closure of some of its steel mills. Currency appreciation has also dented foreign tourist trade. Large expansion of employment opportunities in the mines has led to the shrinking of available worker pool. That is leading to a shortage of workers for other industries and services. Iron ore mining industry is also sucking up a majority of available capital leading to underinvestment in other industries.

9.2.9.2 Longer-Term Negative Consequences

While short-term negative consequences are somewhat counterbalanced by the benign effects of iron ore mining boom, longer-term consequences appear to be more sinister. The boom is the result of the demand for iron ore from a single country—China. What would the effect be on Australian economy as China's industrial growth slows down? In addition, the boom is hollowing out other parts Australian industries and any abrupt change in China's demand will lead to serious economic consequence for whole of Australia.

Perception of an infinite iron ore reserve may lead to complacency on the part of mine owners and those that are benefitted by the boom. Additionally, depending only on selling raw materials is often called a lazy way of making money, which will lead to the decline in available reserves.

9.2.9.3 Reducing Negative Consequences

There are a number of possible actions that can be taken to reduce short- and long-term negative impacts of the iron ore export business. Finding other overseas customers for its iron ores will reduce the effect of any sudden reduction of its demands from China. Establishing steel mills in countries with low labor cost that can supply finished steel at a competitive price to China and other countries can also act as a buffer. Encouraging diversification by offering tax and other incentives to establish and nurture other industries in Australia will go a long way in reducing the effects of sudden reduction in demand for iron ore. Finally, by creating a sovereign wealth or slush fund for use during a downturn will also reduce its impact.

9.3 Minimizing Unintended Consequences

Unintended consequences are a part of our daily life as individuals, as members of organizations, societies, and nations. When our children or we fall ill, we fret about the side effects of taking prescribed medication. When we discipline our children, we worry about that being too little or too much and its consequences. We are now well aware of the results of US government policy of increasing home ownership, which was considered benign, by reducing scrutiny on borrowers' ability to repay mortgage loans. That led to the greatest housing bubble in recent years. Unintended consequence of arming religious zealots to fight against Russian occupation of Afghanistan contributed to the rise of fundamentalist organizations that are now threatening the stability of that and other neighboring countries along with the spread of worldwide terrorist networks, the list goes on and on.

An unintended consequence often happens when a simple system tries to regulate a complex system. A political system is simple as it operates with limited information, short time horizons, low feedback, and poor and misaligned incentives. Social and biological systems, in contrast, are complex, evolving, and with multiple feedbacks. Thus, it is no wonder that many government policies fail to live up to their original promise.

Can system science minimize or prevent unintended consequences? Answer to that question is yes and no, depending on our knowledge of the system in question. Here, it is helpful to quote the well-known words of Donald Rumsfeld, former US Secretary of Defense about things that we know and those that we do not know:

> There are known knowns—things we know we know.
> We also know, there are known unknowns—that is, we know that there are things about which we do not know.
> However, there are also unknown unknowns—those are the ones about which we do not know that we do not know.

Although we may agree or disagree with Donald Rumsfeld regarding the way he conducted the war in Iraq, his basic points on knowns and unknowns are valid and significant from a system perspective.

For known knowns, the answer, to the question on preventing or minimizing unintended consequences, is yes. Here, we can apply system knowledge to create realistic loop diagrams by including all major factors that reinforce and balance the system under consideration. Thus, we will be able to perceive the consequences of various actions, estimate their positive and negative effects, and proceed accordingly. We may go a step further by modeling and simulating the system to find the nature and the weight of unintended consequence and device ways to minimize the negatives. However, the process is iterative, where the models may require updating as new information on the system's behavior become available.

In the case of known unknowns, we can take a similar approach by creating loop diagrams based on inspired guesses for those factors that are unknown. We may take conservative or worst-case scenario approach to estimate their effects. That will result in outcomes that would be better than what mental models can produce. As before, the models would require updating as new information come to light.

In the case of unknown unknowns, their cause and effects and their time of occurrence cannot be predicted. Therefore, there is no way to incorporate them beforehand in loop diagrams or models. However, there are approaches that can be taken to minimize their effects. That is, by tackling them in reactive mode, i.e., taking actions when the unintended consequences start to manifest. They may then be analyzed by drawing appropriate loop diagrams or by updating existing models to reflect the reality.

In the first chapter, there is an example of a bus transportation system, where there are codified procedures for drivers to follow in their normal operations. However, the bus company may also formulate procedures for taking care of known contingency situations, such as:

- A driver reporting sick shortly before his or her starting time
- A bus has a mechanical breakdown while in operation
- A driver experiences heart attack or similar physical ailment while driving.

The bus company may experience known business problems, for which they may have generated sets of coping procedures. However, the full extent or effects of these problems may be unknown, such as:

- Dramatic escalation of diesel fuel price, where the extent of price increase is not known
- Earthquake may cause damages to roads and bridges making them difficult to navigate, while the extent of damage and the time to repair are unknown
- Drivers and other supporting workers may go on strike seeking better pay and benefits, while the extent of their resolve to stick out without getting paid is unknown

For coping with these situations, the bus company may make models based on inspired guesses on the extent of these problems. These models may be updated as and when these problems occur and more data become available.

The company may also face any number of unknown problems in the future about which it can do nothing now. There, the company may formulate its course of actions only after they become apparent. However, agility and flexibility is the key to addressing such problems at their early stages of manifestation for minimizing their negative consequences.

10

The Seven Habits of a System Savvy Person

Abstract A system savvy person is one who lives by the principle that the whole is greater than the sum of its parts. One who believes that the functioning of a part cannot be properly understood without considering its relations with its environment and with other related systems. This is in contrast to the reductionists, who believe that all phenomena can be explained and understood by dissecting them into their smallest components and examining their individual properties. A system savvy person makes realistic models and is better equipped to optimize a system by focusing on the underlying causes rather than on its symptoms. This person strives to make a system more efficient by reducing delays, removing various barriers, and by making continuous improvements.

This concluding chapter highlights seven main system issues, which were covered in earlier chapters. It starts with a discussion on the creation of right mental and conceptual models and then on to building realistic and useful models. That is followed by considerations on optimization, robustness, making continuous improvements, and reducing unintended consequences. The chapter concludes with a discussion on holistic worldview.

10.1 Create Right Mental and Conceptual Models

We all create mental and conceptual models to deal with the systems we encounter in our work and in our daily lives. These models need to be realistic and detailed enough to help us understand the problems and to find their solutions. Often, those models are precursor to more formal models, which are used for simulation

A. Ghosh, *Dynamic Systems for Everyone*,
DOI 10.1007/978-3-319-43943-3_10

and optimization. There are four items that are of prime importance to the creation and use of these models. They are setting perspective at right level, closing feedback loops, thinking dynamically, and documenting them properly.

10.1.1 Set Perspective at the Right Level

During the cholera outbreak in London in nineteenth century, John Snow attempted to understand its cause by focusing on the percentage of population afflicted in each borough of the city rather than on individual cases. In doing so, he was able to identify a public well whose water was contaminated with cholera bacillus. Thus, by setting right perspective, John Snow was able to find the source of outbreak, which had eluded others.

Here, taking the right perspective means looking at the right level for the problem under consideration. A common English expression, "can't see the forest for the trees," aptly describes the predicament. It means that one is unable to see the bigger picture if the focus is on the details. There one needs to step back to get the right perspective.

The level of detail needs to vary according to the nature of the problem that is under consideration. A doctor treating a cholera patient needs to focus on the condition of that patient rather than on the general conditions of a group that are afflicted, whereas, a researcher trying to find a remedy for cholera needs to focus on a group rather than on individual patients.

10.1.2 Close Feedback Loops

The concept of feedback is at the core of a dynamic system. While thinking of taking an action, one need to consider the resultant reactions that may augment (reinforcing loop) or diminish (balancing loop) the results of that action in the long run. Outlawing abortion, for example, can jeopardize the health of a woman who seeks to end her pregnancy by leading her to a back alley operation. A better option for reducing unwanted pregnancies might be to change social norms on casual sex, make birth control pills readily available, and make adoption of unwanted babies easier.

A stricter control of illicit drugs may lead to a higher street price, making the business more profitable for both traffickers and dealers, whereas, changing social attitudes on recreational use of illicit drugs and treating addiction as a medical problem could produce better results. In the USA, improving school system leads to increased property value in a town. That leads to increased revenue, thus, making more money available for further educational improvements, which is a reinforcing loop with benign effects.

A person with system knowledge is able to find such interdependencies and by creating loop diagrams is able to augment his or her thought process. That often leads to the discovery of other such connections. Additionally, loop diagrams help in getting the points across to others, which is often difficult with mere words. A word of caution—there is always the possibility of going overboard with many feedback loops jumbled together in one model leading to an incomprehensible situation. That may be avoided by eliminating those feedbacks that have little effect on system functions that are under consideration. Additionally, complexity may be reduced by drawing several loop diagrams, one for each subsystem.

10.1.3 Think Dynamically

When considering historical events, such as the cause of Second World War, one may focus on the actions of individuals, such as Hitler, Stalin, and Chamberlain, but that will miss the point. A more appropriate model would be to consider the reasons for the rise of Nazism and Hitler over time after the First World War and the state of the economy and the perceptions of the common people in Germany during that period. Dynamic thinking requires one to look at a problem from the perspective of past, present, and future—how the problem got to that state and where it is going, rather than merely looking at it in a narrow focus.

Deepwater Horizon oil spill, better known as the BP (formerly British Petroleum) oil spill, in the Gulf of Mexico in 2010, which is considered to be the largest accidental marine oil spill in the history of petroleum industry, was caused by an explosion and sinking of the Deepwater Horizon oil rig. Here one may easily get bogged down by focusing on details rather than on the dynamics of the system that led to this catastrophe. A report pointed to the defective cement on the well, faulting mainly BP, but also rig operator Transocean, and the contractor Halliburton. However, the underlying cause, according to a White House commission, lay in a series of cost-cutting measures that were taken by BP over a period of time, which led to insufficient safety systems in place. Thus, the spill was the result of a set of systemic problem, which would recur if not addressed properly.

10.1.4 Document the Models

We human beings are creating mental models all the time; however, they are limited by our brain capacity. We frequently fail to create a model that is accurate enough when the number of variables in a system is more than a limited few and their relationships are nonlinear. Additionally, most of us do not have a photographic memory to remember a model thus created accurately and in its entirety. We also find it difficult to convey such a mental model accurately to another person. Therefore, there is the need to document a model in an electronic format or on paper.

Documentation formats vary from narratives to block diagrams, various forms of loop diagrams, and mathematical equations. The most common forms of documenting a system are control loop diagrams for engineered systems and causal loop diagrams for all others. Modeling and simulation packages often provide documentation facilities, and there are software packages that are meant exclusively for that purpose.

10.2 Make Interactive Models That Are Realistic and Useful

With an interactive model, one is able to alter its variables to simulate various scenarios, making it a valuable tool for learning, research, and problem solving. Today there are many cost effective and free software packages available for modeling and simulation on personal computers. They do not require deep mathematical knowledge and are thus suitable for experts in many different disciplines.

A model could be designed for a system that is yet to be built or could be for one that is already in operation. For a system that is already built, it is helpful to know the functions of its various subsystems and their interactions. A model builder needs to have a clear idea of the model's purpose and scope. Then a model may be built based on the available knowledge of the system, which may be refined as more information becomes available as the model building progresses. Black box modeling technique like artificial neural network may be used if little or no information is available about the inner workings of the system.

10.2.1 Choose Right Modeling Technique

Consider a system like the predator and prey relationships of deer and wolves. Here, we may put all the deer in one group and all the wolves in another, as we are not interested in the fate of individual deer or the actions of individual wolves. Similar is the case with a business, where one wants to track the changing buying habits of a group of customers. However, the problem is quite different when one wants to track the buying habits of individual customer. In the first case, it would be appropriate to use a conventional modeling package, with stocks and flows, which can simulate a limited number of groups. Whereas, agent-based modeling would allow the tracking of individual behavior patterns of a large number of entities.

Therefore, before building a model, we need to decide whether to look at the system as a set of groups or a collection of individuals. That of course depends on the problem we are trying to understand and solve. It is conceptually easier to lump individuals into groups, and if the number of groups were small, it would be less arduous to build a model in a conventional way, with stocks and flows.

Configuring an agent-based model requires much more effort than for a group-based model as each individual's unique characteristics need to be specified. In addition, more time and computer power are needed for its simulation. So, one should consider a system problem carefully and weigh the advantages and disadvantages of conventional versus agent-based modeling and simulation techniques before proceeding to build one.

10.2.2 Recognize Nonlinearity and Time Variance

In a nonlinear system, the outputs are not always proportional to its inputs. For example, the amount of effort needed to cook a meal for four people is generally less than twice the effort it takes to cook the same meal for two. Thus, the relationship between cooking effort and the number of people served is not linear. Such nonlinearities are common in nature, like the height of a tree or the circumference of its trunk is related to its age, but they are not linearly proportional.

Nonlinearity increases the behavioral complexity of a system and bedevils most people. When nonlinear relationships are well defined, it is usually possible to create an accurate model, as in the case of earth's gravity. However, making accurate models gets difficult when nonlinear relationships are not well understood. A system expert is usually able to identify the nonlinearities and is, thus, able to estimate the accuracy of the results of a simulation. In order to reduce complexity, the system expert may simulate a nonlinear relationship for a limited region, where the relationship is approximately linear.

The behavior of engineered systems is generally time invariant, that is they change little with time, which is not the case for many natural and social systems. An automobile, for example, is expected to behave in the same way today and tomorrow as it did yesterday, unless there is a degradation or failure of one or more of its components. Whereas, consumer tastes vary with time, what was trendy last year may not be so today. A seasoned marketing person understands such variations and makes frequent surveys to keep up with the change.

10.2.3 Recognize the Power and Limitations of a Model

Building a model and observing its behavior is an effective way to understand a system. An interactive model is worth a thousand pictures. Often a model is the only viable way to extrapolate the behavior patterns of a large or a complex system. However, one needs to understand a model's essential limitation; that is, it is not the actual system, but an abstraction of some of its essential parts. In addition, a model is only as good as the variables and their relational values used in building it. A model does not give true pictures when they are incomplete or inaccurate.

In many systems, the initial or starting conditions affect their behavior significantly. If those conditions are not correctly specified at the start of a simulation process, the model produces erroneous results. Moreover, there are limits to the values of input variables for all models. Beyond those limits, the behavior of a system may be incorrect or impossible to predict. Thus, when using a model, the user needs to understand and appreciate both its capabilities and its limitations.

10.3 Optimize the System

A system functions better when it is optimized but there may be several barriers. Not paying enough attention to these barriers often leads to frustrations and failures in achieving the desired goals. The impediments to optimization and their intensities vary between systems. While some of these barriers may be removed or modified easily, there are others that are tough or insurmountable. Some of these barriers are discussed here.

10.3.1 Minimize Delay and Dead Time

When one pushes the gas pedal in a car to increase its speed, the result is not immediate. The actual time it takes a car to reach the desired speed depends largely on its weight and its engine size. Higher the power of the engine, shorter is the time for the car to get to the desired speed, hence the attractiveness of high-powered cars.

Delay and dead time are major barriers to the efficient working of a system. It is easy to get frustrated when one acts, but does not get the desired results immediately. There can be serious implications in health care, for example, where saving a life may depend on the timeliness in getting an ambulance to a person suffering from a heart attack.

Thus, reducing delay or dead time in a system is one of the most effective ways to increase its efficiency. A system designer needs to take that into account while designing a system. There are of course inherent delays due to the natural laws that are impossible to overcome, but closer examination will show that there is room for improvements in many situations.

10.3.2 Address System Constraints

A personnel manager faces constraints in hiring right persons for his company. That may be due to external factors like the state of the economy or the right educational curriculum in the local colleges. It may also be due to one or more internal factors, such as the salary scale offered or the reputation of the company as an employer. Constraints such as these are ubiquitous; they are present in every system. For example, in a railway system, the speed of a train may be constrained by the power of the engine, but more often, it is the condition of the track that sets the limit.

Internal constraints are often easier to fix than external factors. Trying to fix constraints that are of lesser importance or those over which one has little control often produce ineffective or temporary results. A seasoned system practitioner identifies the main constraints in a system, and thus able to address them effectively.

10.3.3 Focus on the Underlying Causes Rather Than on Symptoms

Conventional medicine is focused on acute care, not chronic illness where doctors often treat symptoms rather than root causes of a malady. Many disorders have multiple underlying factors that take time, trial and error, and deep thinking to figure them out. Why does the patient have a disorder, what are the underlying causes that can be addressed? Such a thought process should be there at the start of a diagnosis.

Today, most doctors simply do not have time to find those root causes, which is leading many patients to seek alternate medicines. This state of affairs may be addressed by taking a system-oriented view, where the emphasis is shifted in training of the primary care physicians from treating a malady to taking care of the whole body, both physical and mental. In addition, the method of payments from insurance providers needs to be changed so that doctors are reimbursed based on keeping their patients healthy rather than on the number of visits, the patients make.

10.3.4 Remove Obstacles Rather Than Push Harder

A person who starts a fasting regime to reduce his body fat may experience some initial success. However, he may soon find that satisfying his increased craving for fatty foods on non-fasting days largely nullifies his fasting efforts. Similarly, a social activist often gets frustrated when his efforts to improve a certain aspect of a society produce little long-term effects. For example, building a modern medical facility in an underdeveloped region often has limited effect unless there are corresponding improvements in related areas, such as sanitation, supply of clean water, and availability of trained personnel. A system tends to drift back to its original state soon after the initial efforts for change are completed. Various forces, such as a traditional mindset, lack of proper infrastructure, or lack of training to maintain the improvements, bring the system back near to its original state.

That generally happens because for most of the time, a system is in a balanced state, where positive and negative feedbacks acting on the system balance each other to keep the system in the way it is. An external signal may alter a system, but its effects may be minimal or transitory because the balancing loops tend to bring the system back to its original state.

Social and cultural norms are often significant barriers in human societies. They may take the form of trade barriers, restrictions on free exchange of ideas, "not invented here" syndrome, and many others. Here, weakening the negative feedbacks, that are holding the progress back, may be more effective than augmenting the positive feedbacks.

10.4 Make Systems More Robust

Robustness is the ability of a system to withstand severe blows while still carrying on its expected functions. Most biological systems are robust; otherwise, they would not have survived in their long evolutionary history. We human beings are able to fend off the onslaughts of various germs most of the time. Trees in their natural habitats survive vagaries of weather, such as wind, snowstorms, and drought conditions. This robustness is due to the many feedback loops that are able to restore these systems even after large disturbances. Engineered, social, political, and economic systems are not always as robust the natural ones. However, a long-surviving human institution is usually more robust than one that has been created recently because the former has coped with many small and large changes over time.

There are many attributes like stability, decentralization, and redundancy that make a system more robust. While designing a new system it is the designers' responsibility to incorporate them to increase robustness.

10.4.1 Minimize Uncontrolled Oscillations and Escalations

Most systems contain reinforcing and balancing loops along with delays. In the housing market, for example, significant delays between planning, construction, and sale of a property make it difficult for builders to adjust their schedules quickly. Overinvestment in housing market during rapid rise of real estate price leads to a housing bubble and its subsequent collapse. There are many other examples of such oscillation, including the number of predators and prey in a natural habitat, and the cycles in national and world economies.

One way to reduce such oscillations is to reduce the gain in the system, like a homebuilder setting a self-imposed quota. However, fine-tuning of the system is necessary because too much conservatism will lead to a sluggish response, which will depress the profits of a homebuilder and slow the growth of the economy to an unacceptable level.

Delay between actions and their effects should be reduced wherever possible to reduce such oscillations. "Just in time" manufacturing practice is certainly a step in the right direction. However, for homebuilders, for those who control national economies, and for a host of others, system delays are inevitable and cannot be eliminated. Better forecasting of market demands or national economies is of great help, but their predictions are often not accurate enough due to unforeseen external events.

Technological advancements are reducing delays, by giving near instant communications between individuals, businesses, and nations. Publishers are now able to print a single copy of a book on demand rather than a thousand copies in one go, thus reducing warehousing cost and wastage.

In a social setting, differences in opinion can easily escalate into verbal arguments and even to physical encounters. This is a high gain closed loop situation, where difference between expectation and feedback is amplified at successive iteration of vocal exchanges between individuals and parties. One or more person in the group may diffuse such a situation quickly by attenuating their response, thus, reducing the gain in the system.

Similar is the case of two antagonistic nations, where increased rhetoric can quickly lead to a break in their relationship and the start of hostilities. Seasoned diplomats with system expertise are often able to bring such situations under control by just reducing their rhetoric.

10.4.2 Design Decentralized Systems

Decentralization increases robustness, where the failure of a part of a system does not make the whole system inoperative. Internet is a prime example of such a decentralized architecture. However, decentralization does not mean a total freedom that allows subsystems to operate in any way they like. Rather, the subsystems should be designed to function in cooperative fashion to fulfill commonly agreed mission while staying independent to take care of local variations and needs. A geographically distributed workforce linked by telephone and Internet communications is less vulnerable to a catastrophic incident that can destroy a single facility. A multinational manufacturing company needs to be decentralized to take care of the variation of the labor supply, market conditions, and the law of the countries it serves while adhering to its core principles of quality and service.

10.4.3 Promote Redundancy

Redundancy is a major factor for increasing the robustness of a system. An engineered system like a large airplane with two or more engines is designed to fly safely even when one of its engines fails. Today, most large aircraft are fitted with autopilots where computers can take over some of the human functions. There, "a multiple redundant" architecture is used allowing a standby computer to take over the autopilot function from one that fails. Likewise, many automobiles are now fitted with dual-braking system. Similar is the case with other large systems where safety is of prime importance, such as in oil refineries and nuclear power plants, where their safety and control functions have dual, triple, or even higher levels of redundancies.

Examples of redundancy and decentralization abound. A good business leader gives due importance to training his or her successors, thus increasing redundancy in the system. A good sales manager augments decentralization by allowing the development strategies at local levels to take care of market variations. In a healthy family, parents give their children enough freedom to operate independently, based on their age and maturity, while instilling in them the core values, such as honesty, discipline, and hard work.

Redundancy and decentralization are quite common in biological systems though it varies between species. A tree does not die when one of its branches is damaged or severed. We can survive small cuts and bruises, where damaged cells are quickly replaced by healthy ones. Most animals, including humans, can survive damage to some of their limbs and some animals are even able to regrow them when broken or severed.

A democracy with federal, state, and local authorities functions as a multilayered decentralized system. When a leader dies or is considered unable to function, there is usually a pool of suitably qualified persons available to take his or her position. That is not the case with an authoritarian or dictatorial system, where the demise of a leader often leads to a power vacuum.

10.5 Strive for Continuous Improvements

Most of the time we do not tinker with systems we encounter in our daily lives, so long they function satisfactorily. "If it ain't broke, don't fix it" is the mantra. Thus, we normally do not make any conscious effort to improve a system until it becomes dysfunctional, significantly degraded, or has broken down completely. However, a system-oriented person understands that most systems, as good as they are, can often be improved by conscious efforts. To improve a system one may follow Lao Tzu, the ancient Chinese philosopher, who said, "A journey of a thousand miles begins with the first step." The technique of taking small steps can begin with something as simple as making a commitment to get more fit by doing physical exercise starting today. This week it will be 2 min a day, next week it will 4 min a day, and the like until the goal of 20 min a day is reached. As the exercise time increases in small increments, one may not notice the extra effort put into making them, and will not get discouraged easily. Thus, one is able to reach the goal more easily with lesser chances of failure.

A step-by-step approach works well in many other areas, such as in building relationships between individuals, groups, and nations. An example is the ping-pong diplomacy and its follow-up that thawed the antagonistic relationship between China and the USA in the early days of cold war. Thus, one can always find ways to make continuous improvements, albeit in small steps to make a system more useful and efficient.

On striving for continuous improvements, one cannot preclude the possibility of paradigm shifts or revolutionary changes that happen from time to time due to advancements in science, technology, or in our social and political systems. They need to be embraced if they are for the common good. However, such changes do not mean the end of improvements in small steps. There is always room for further improvements even after a big change.

A systematic approach to improve the quality of a product, often called Kaizen, was spearheaded by Toyota in Japan. There, a worker is expected to stop his moving production lines in case of an abnormality and, along with his supervisor, suggest improvements to resolve the problem. In addition, meetings of workers and managers are held at regular intervals to address manufacturing problems and to improve production process. Kaizen has vastly improved the quality and reliability of cars manufactured in Japan and the practice has since been adopted by other industries in Japan and in many other countries.

10.6 Reduce Unintended Consequences

Before 1935, Australia did not have any toad species of its own. However, the country did have a major beetle problem. The beetle's larvae were eating the roots of sugarcane and stunting, if not killing, these plants. The proposed solution to this problem was to import cane toads from Hawaii.

The plan backfired completely as it turned out that cane toads could not jump very high, so they were unable to catch the beetles that lived in the upper stalks of sugarcanes. Instead of going after the beetles, the cane toads began going after other species, such as insects, bird's eggs, and native frogs. As the toads are poisonous, they began to kill would-be predators thus causing immense damage to native species. Fish died by eating the tadpoles and other animals died by eating adult toads. This is a perfect example of unintended consequences, which is largely due to the lack of necessary system study on the possible effects of introducing a species in a new environment.

In recent years, the Australian government has embarked on a multi-million dollar campaign to eradicate the toads that have spread over a major part of Australia. Various methods of eradication have been proposed, such as specially designed traps to catch them without luring the native frogs and specially designed virus that will kill them without harming other species.

Invasive species are a major problem around the globe; arriving in new ecosystems where there are few natural predators, they wreak environmental havoc and often cause considerable economic damages. There are of course many other examples, like the proliferation of kudzu vine in southern United States, which was imported from Japan to control soil erosion. When left uncontrolled, kudzu proliferates fast, growing over almost any fixed object in its proximity including other vegetation. It smothers other plant species and over a period, it can kill a tree by blocking the sunlight.

There was mass starvation in China during the Great Leap Forward when Mao Tse-tung ordered that all sparrows be culled as they fed on grain seeds planted on the fields. However, doing so meant that there were no birds left to eat the locusts that fed on the crops. That led locust swarms to take over the country, devouring entire crops.

More recently, there has been the rise of ISIS in the Middle East, which can be attributed to the removal of Saddam Hussein causing political vacuum and anarchy in Iraq. Many other examples of unintended consequences are described in an earlier chapter of this book.

Some experts are of the opinion that there are no side effects or unintended consequences, just effects. Those that prove beneficial are called main effects and we claim credit for our actions. Those that are detrimental are termed side effects or unintended consequences thus hoping to overlook the failure of our interventions. There is some truth in this, as it is often impossible to anticipate all possible feedback paths and their weights beforehand. That is particularly true for problems that are new, with limited or no prior history. There, one needs to address the emerging side effects at their earliest possible occurrence, which may considerably reduce their negative consequences. Thus, while it is not always possible to eliminate unintended consequence, by doing proper system analysis one can often minimize its occurrence or reduce its adverse effects.

10.7 Cultivate a Holistic World View

Western scientific method has been dominated by specialization and reductionism. There the idea is that all natural phenomena can be explained and understood in terms of their smallest parts. These ideas have been highly successful and have led to a great many major scientific breakthroughs, such as discovery of medicines for curing various illnesses. However, they often do not take into account their effects on a patient's health and general wellbeing. A miracle drug often does not live up to its expectations because of its negative side effects, or because the bacteria it is supposed to kill starts developing immunity. Thus, this approach is now increasingly being called into question.

Holism, which is synonymous with system view, holds that the whole is greater than the sum of its parts. It asserts that the parts cannot be properly understood except in their relation to the whole. Rather than trying to find a single "right" answer, holism focuses on finding balanced answers that address seemingly contradictory goals like efficiency and resilience, collaboration and competition, aesthetics and utility, and diversity and consistency. That leads to a deeper understanding, where one recognizes the fundamental interconnectedness of all phenomena and the fact that, as individuals and societies, we all are dependent on each other and on our common natural resources, such as air, water, and energy.

Holistic science takes a broad, integrated, systemic approach to many fields, including environmental management, sustainability, biological research, medical science, public health, and business management. In many countries, the newer generations—the post baby boomers are more aware of the perils of uncontrolled developments that strain the societies by increasing pollution and greenhouse gases. They want holistic medical therapy where the health of the whole body and mind is given due consideration along with localized treatment of an illness.

More than 2500 years back, Gautama Buddha taught that all phenomena are interrelated. He asserted that nothing exists in isolation. In other words, all beings and phenomena exist or occur only because of their relationship with others. Nothing can exist in absolute independence of other things or arise of its own accord. That is the holistic worldview, where one sees the world as an integrated whole rather than a disassociated collection of parts. That leads to an ecological awareness, and an understanding that we are all interconnected and ultimately dependent on each other. We need to cultivate a holistic outlook to be in a better position to effectively tackle many of the professional, social, and personal problems that we all encounter in today's world.

Epilogue: Role of System Science in the Twenty-First Century

Abstract We are now witnessing major technological advancements in areas, like artificial intelligence, robotics, and self-driven cars. In biological field, we are witnessing medical breakthroughs based on the ability to look inside a brain, genome mapping, and the symbiosis between microprocessors and biological entities. The pace of change is accelerating, as the world around us is getting more complex. These changes are affecting all human societies and challenging us with new and complex technological and social problems that can only be addressed effectively by system knowledge and expertise.

This section outlines some the opportunities and challenges associated with this revolution. It discusses the formidable technical challenges in designing a self-driven car and the resulting societal challenges we will face when they are available to the general public. It outlines the role of personal robots, in helping elderly and disabled. How Internet of Things, which will soon be pervasive, will help in communicating between man and machine and between mechanical entities themselves. How functional MRI will open new vistas in understanding brain functions leading to their better regulations and thus help in curing mental problems. Finally, it outlines ways to impart system knowledge starting from schoolchildren to adults.

E1 The Fourth Industrial Revolution

Humanity is poised for another industrial as well as social revolution. It is a paradigm shift rather than incremental change, which will cause discontinuities in the way human societies operate. Technological changes include the

A. Ghosh, *Dynamic Systems for Everyone*,
DOI 10.1007/978-3-319-43943-3

way we communicate, how we manufacture goods for our material needs, and services we provide to supply them to individual citizens. The rate of these changes will accelerate in the first half of this century. However, the resultant and necessary societal changes will not happen that fast. That will create a lot of social tensions and problems, which need to be addressed.

The First Industrial Revolution, which started in the late eighteenth century, was a shift from the reliance on animal and human labors to mechanical power produced by burning fossil fuels. The Second Industrial Revolution, which started in the late nineteenth century, brought major breakthroughs in electric power generation and distribution, mass production, and wired and wireless communications. The Third Industrial Revolution began in mid-twentieth century with the development of digital systems for computation and communication, that led the way to automation of many production processes and faster processing and sharing of information.

We are now witnessing major changes that include dramatic increase in the capabilities of artificial intelligence and robotics, development of self-driven cars, near universal availability of Internet, 3D printing as a viable manufacturing process, and the applications of nanotechnology. In biological field, we are witnessing medical breakthroughs based on the ability to look inside a brain, genome mapping, and the symbiosis between microprocessors and biological entities. The pace of change is accelerating and it is difficult to envision all the major changes that may take place in near future. However, it is certain that these changes will fundamentally affect all individuals and societies.

Some detractors are saying that the Fourth Industrial Revolution is merely an extension of the third, however, the rate of technological change and resultant effects on human societies will be much more profound and dramatic in this era, than has ever been witnessed in the past. Thus, it may aptly be termed as another revolution.

The Systems Approach

According to many experts the three most important skills that will be needed in this Fourth Industrial Revolution are Complex Problem Solving, Critical Thinking, and Creativity. System expertise is of course critical for complex problem solving. It offers a broader perspective to those who are in the cutting edge of the new technologies, such as designing driverless cars, intelligent robots, or interfacing brain with external device. It is also crucial for the techno-sociologist and politicians that will solve the social problems resulting from these technological advancements.

E1.1 Opportunities and Challenges

Currently, more than a third of the populations in the world are using social media platforms to connect, learn, and share information, which will significantly increase in the coming years. Products based on digital technology, such as computers and cell phones, are already producing beneficial effects for a large number of people in many countries. Technological innovations are producing gains in efficiency and productivity resulting in the reduction of cost of goods and services for a large section of the population. Thus, the Fourth Industrial Revolution has the potential to raise living standards and improve quality of life of people all around the world. However, a society's adjustment to technical advancements is often slow and difficult. It took almost 50 years for electric power to become commonly used after its introduction and personal computers took almost a decade to get accepted by the general public.

Automation in a much wider scale will cause downward pressure on wages for low skilled workers. It will also affect the employment prospects of those with middle skills. In contrast, persons with higher skills especially those in the cutting edges of technology will be in high demand. That will create a greater divide between persons with high skill and education and the rest of the society.

That may also lead to a premature deindustrialization in many developing countries. Governments in those countries will face serious unemployment problem because they will not be able to find work for the unskilled labor from rural areas. As the precise course of these rapid changes is difficult to predict business leaders will have to repeatedly reexamine their strategies to keep their business solvent and to keep them growing.

If the governments and their societies make right moves, they will be able to channel the technological changes that will benefit a broader spectrum of the societies. If they do not act or go in wrong directions, they may have to deal with both angry unemployed workers and resentful taxpayers. In the following sections, we will discuss some of the technical advances that are now taking place along with the role system knowledge will play in coping with these changes.

E2 Self-driven Cars

Smith family has recently sold their last car. A self-driven car now takes Brad Smith to his workplace every morning along with three of his co-workers and returns them back after work. Brad and his colleagues have formed a co-op, which owns the car. The car is electrically operated and is fully self-driven, with its preset start, destination, and intermediate stopping points. Those locations can, however, be changed from a handheld cell phone with appropriate code. In the evening, the car parks itself in a pre-assigned spot and gets it charged for the next day. On weekends, Brad may book the car for a few hours to shop or to run errands.

Brad's son Kyle who is in Junior High goes to school on a driverless bus along with a teachers' aide who travels with the students to look after them and to maintain order. Brad's wife Brenda, who works part-time, summons a driverless cab from her cell phone for going and returning from work. Their daughter Brigitte, who is in college, usually cycles around the campus. When she comes home, which is around once a month, she takes a commuter train to the station closest to home, and a driverless cab waits for her at the station.

Last winter, the Smith family rented a driverless mini-van with sleeping facilities for a vacation to Florida. At night the family slept, while the van drove on, thus significantly reducing their stress and travel time. The Smith family is now saving a lot of money by not owning the two cars they had before along with the hassle of driving to the city and finding places to park. In his youth, Brad loved driving and often tinkered with his car but his garage is now empty. He is thinking of getting a vintage car that he can play with and take it to auto shows. After all, manually driven cars are going to be vintage fairly soon.

Welcome to the era of driverless cars. This era has not fully dawned yet but will be happening in a not too distant future, possibly sooner than we anticipate. Google has a prototype version, which has made trial runs on the streets of San Francisco. Traditional automobile manufacturers and others, such as Apple, are also getting into the act. It is a fast evolving technology, which only a few years back was considered to be in the realm of science fiction. Rapid advancements have taken place since DARPA's Grand Challenges in the years 2004, 2005, and 2007. These challenges were designed to spur the development of technologies needed to create fully autonomous ground vehicles. Still there are huge problems in designing and building a fully functional self-driven car, at a reasonable cost, that may operate on public streets, without manual backup. Such a car needs to be able to cope with almost all possible contingencies, including snow, fog, and rain, however slight their chances may be.

E2.1 The Challenges

Designing a fully functional self-driven car is a major challenge for system designers, where safe and effective control depends on a large number variable that need to be monitored and controlled. The functions of self-driven cars may be considered as extensions of some of the existing features that are available today, such as navigation aid, adaptive cruise control, auto parking, and emergency braking. While these features are helpful, it is assumed that the backup of human drivers is readily available when they malfunction. However, managing human behavior behind the wheel of a fully self-driving car may be very problematic because human beings generally do a poor job of monitoring automated systems, and so they rarely are able to retake control quickly enough, if needed.

A self-driven vehicle needs to know its exact geographical location along with the positions of objects such as road edges, other cars, pedestrians, and current road conditions (Fig. E1). Now GPS satellite technology can determine geographic locations accurately for most of the time, but can be unreliable in some urban streets where high-rise buildings may block the satellite signals. Moreover, the standard GPS is not precise enough to control a vehicle on a busy street. Thus, additional sensing devices are needed to reduce positioning error down to a few inches.

Additionally, detailed geographical maps are needed for information, such as road topology and the locations of traffic lights. The system needs to be foolproof, tamper resistant, fault tolerant, and reliable. Thus, designing a fully functional self-driven car that can safely drive on public roads without any manual intervention both in normal conditions and for all emergencies is a challenge of a much higher magnitude. There are also economic, social, legal, and other issues (detailed later) that need to be solved to put a driverless vehicle on public roads.

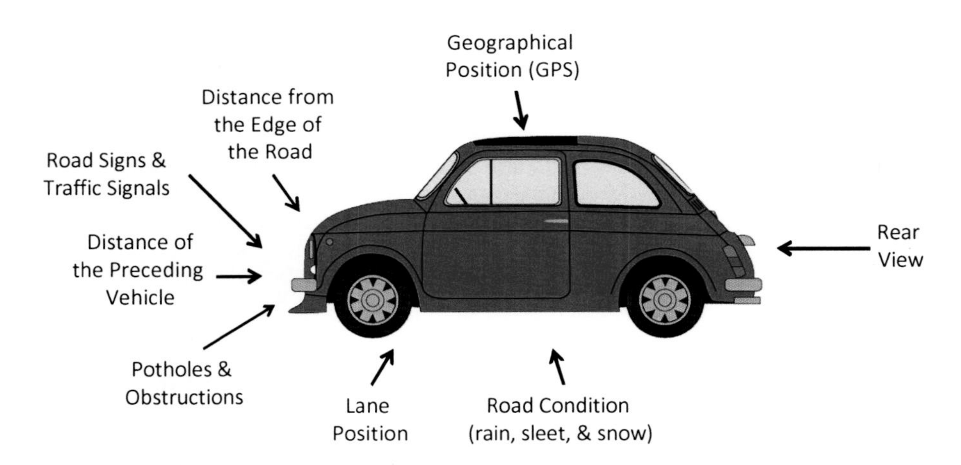

Fig. E1 Major feedbacks received by a self-driven car

E2.2 Artificial Intelligence

Artificial intelligence (AI) is a key component for analyzing the feedback information for controlling a self-driven vehicle. Data acquired with the help of various sensors need to be evaluated every split second to determine how fast a car should run, accelerate, slow down, or stop, and when to turn or reverse. The goal is to take the car to the destination both safely and legally. Additionally, these cars should be able to communicate with each other to warn on possible dangers or traffic congestions. Thus, the value of a driverless car is now tilting from its physical hardware to sensors and on-board artificial intelligence software.

One of the busiest research areas these days is about teaching machines to truly see and comprehend their surroundings. Massively complex neural network systems are being built and tested in various research laboratories. Such a system analyzes live video feed from a vehicle's camera and instantly sorts objects from the field of view into many different categories, such as road, building, vehicle, pedestrian, bike, tree, or road sign. Currently such a system can label more than 90 % of the objects correctly and there is ongoing research on increasing their accuracy. As a human driver gets more efficient in driving on an oft-travelled route, a driverless car will also be able to do the same with its self-learning capabilities. Thus, it will be able to remember road conditions, location of traffic signals, and other landmarks for a travelled route, making it more efficient in reaching its destination in subsequent journeys.

The Importance of System Science

A self-driven car incorporates a number of complex systems, such as vision, road positioning, navigation, safety, and communication. Each of them requires multiple feedback loops that need to be designed and accurately tuned. Additionally, these systems need to coordinate optimally, increasing the complexity of the total system by orders of magnitude. Experts are of the opinion that the navigational complexity of a self-driven car far exceeds that of an autopilot of a modern jet liner. The system also needs to be cost effective to build and operate; otherwise, it will not be acceptable to the public. Thus, system expertise plays a major role in its overall design and in the design of its individual components.

In addition to the technical challenges, there are a number of social and legal issues that need to be addressed to deal with this disruptive technology. They include legal liabilities in case of an accident, and the needed improvements and alterations to roads, lane markings, traffic signals, and road signs. Urban planning will be affected because there will be fewer cars, which will reduce the need for parking spaces in residential and shopping areas. Automobile manufacturers and the society in general will need to address social changes for making self-driven vehicles acceptable to the taxicab operators, trucking companies, and the public. Here, sociologists with deep expertise will be needed to generate computer models to study many different scenarios before implementing the necessary changes.

E2.3 Payoffs and Benefits

Currently available automotive innovations, such as adaptive cruise control, auto parking, and emergency braking, have resulted in increasing safety but offer limited additional benefit. Fully automated driverless cars will provide increased safety and unleash the driver from the steering wheel, thereby increasing his or her available time, which is economically precious and a scarce resource. The benefits are most obvious for driverless trucks and vans, where the expenses on manual drivers represent about a third of the total transportation cost. That will lower the cost of consumer goods, thus boosting the economy of a country.

Fully autonomous cars can be used as self-driving taxis. In urban areas, fleets of such taxis will provide mobility as a service at a much lower cost than privately owned vehicles as their utilization rate will be much higher. Thus, these fleets will be less sensitive towards any additional cost that autonomous vehicles entail and will adopt the technology much earlier than private owners. As the self-driving vehicles are expected to be much safer, there would be a reduction of their insurance premium. However, that may take a few years to take effect after they are introduced.

E2.4 Social Implications

There are major social implications of driverless vehicles. They will increase the mobility of elderly persons, who are unwilling or unable to drive. Young and inexperienced drivers will be able to use them without major insurance penalties. Overnight journeys will become more common if people are able to rest and sleep in their car while traveling, which may also reduce road congestion.

Self-driven cars may lead to a large expansion of ride-sharing services and car clubs, where vehicles owned jointly by its members may become common. As a result, a large percentage of population may not be inclined to own a car, which will lead to the shrinking of car market in advanced economies. That will also significantly reduce the need for garage and parking spaces for residential housing and in shopping areas. The economic and social potential of driverless cars will possibly be enormous and could be disruptive in many ways.

There are also local, national, and international legal issues that need to sorted out by respective authorities. It is an open question as to who would be liable in an accident; is it the owner of the car, the vehicle manufacturer, or the company that supplied the software? Most of the auto industry is treating the problems of self-driven cars as technology issues, but the human and social issues would be much tougher to solve. In many countries and regions, the biggest hurdle will be to convince local and state authorities that fully driverless vehicles are safe on their streets.

E2.5 Meeting the Challenges and the Progress So Far

Google is the leading exponent of fully autonomous motor vehicles. In 2014, Google unveiled a self-driven car with no steering wheel or pedals, which is still in a testing phase. The car consists of two seats and space enough to carry a small amount of luggage. There is no manual control option in this car other than a start button and an emergency stop button.

The car is powered by an electric motor and has a range of around 100 miles when fully charged. It uses a combination of sensors as well as software that utilize Google maps to determine its location in the real world. For navigation, it uses a Global Positioning System (GPS), radar, lasers, and cameras to get accurate pictures of the world around it. It is able to recognize obstacles, pedestrians, traffic lights, other vehicles, and other unpredictable hazardous conditions. The maximum speed of this car is at present limited to 25 miles per hour. Google expects that it will take about 5 years of testing and upgrades before it can offer it to the public. Meanwhile, Google is trying to seek partnership with other traditional car manufacturers like Toyota and Ford.

In addition to Google, Apple, Tesla, and many of the established car manufacturers are busy designing driverless cars. There is a huge competition between them to be the first in offering them to consumers. While partially automated vehicles are available now, the availability of fully automated driverless cars is estimated to be approximately a decade away. Thus, the automotive industry will encounter more changes in the next 10 years than it has experienced in the last 50.

E3 Robotic Systems

A robot is a highly integrated system comprising of mechanical parts, sensors, computers, and software. A robot with reasoning capabilities can act based on signals it gets from its own environment. Growth in the power of silicon chips, digital sensors, speed of communication, and greater applications of system knowledge is improving the performances of these robots. It will increasingly relieve human beings from working in hazardous operations such as remote welding, plant dismantling, fire fighting, and disposing bombs.

Aerial robots, commonly known as drones, are replacing manned aircrafts to fight wars. As they become more cost effective they will increasingly be used for more peaceful purposes, such as in farming, monitoring traffic, fighting forest fires, and much more. Amazon is using robots to automate its warehouses, and is looking into using drones to make home deliveries.

Robots are extensively used in certain industries, such as auto body manufacturing, where they perform routine repetitive tasks. There are also robotic vacuum cleaners that can clean a room without bumping into various objects scattered on the floor. These specialized robots are capable of doing a single or similar type of tasks. Today, the challenge is to build robots that are intelligent and versatile enough to perform a number of different tasks in optimal fashion based on human commands and on environmental signals. These intelligent robots need learning capabilities so they can perform tasks more efficiently with increased experience. Additionally, robots that are designed for domestic use need to be affordable to average householders.

E3.1 Artificial Intelligence and System Science in Robotics

A robot for general use must possess intelligence to do meaningful work for its owners while navigating its surroundings and preventing itself from doing anything that may be harmful or dangerous. Moreover, a robot has to be efficient so that it can operate at a reasonable speed, while still being easy to maintain and modify. Thus, a typical robot needs multiple sensors and microprocessors and be able to be connected to a local area network (LAN). System science plays a very important role in such a design, where hardware, software, and man–machine interactions come together. In addition, system knowledge is essential in the efficient use of multiple robots in an institutional setting like in a battlefield, a hospital, a nursing home, or in an office.

E3.2 Personal Robots for the Elderly

As the elderly population increase, there are greater and more pressing needs of health care and personal assistance at the society and family level. As more people live to older ages, they also face more chronic and limiting illnesses like arthritis, diabetes, osteoporosis, and dementia. Elderly persons are becoming increasingly dependent on others for help in performing their daily activities.

These needs call for a much faster advancement of service robots that can assist elderly people to perform activities such as bathing, getting around inside their homes, and preparing and serving meals. Additionally, the elderly often endure loneliness and/or loss of independence when living in nursing homes or other assisted living facilities. Here, personal robots will be of great help by allowing greater independence to the elderly and infirm and the ability to stay in their own homes with reduced manual supervision and care.

Today, mobile robots can take hospital trolleys where they are needed, taking over much of the work that porters do today. Soon they may replace nurses by performing jobs like dispensing drugs, taking temperatures, and cleaning up wards, thus reducing the cost of caring.

Robots are still inferior to biological systems in terms of dexterity, coordination, and lifting capacity. However, as robotic science quickly progresses, robots will increase their capabilities in the near future to include:

- Improved communications with human beings, not only vocally but also by body language and other non-verbal clues
- Increased learning and reasoning capabilities
- The ability to tap Internet for information, when it encounters a new situation about which it has little knowledge.

E3.3 Robots as Pets

University researchers and tech companies are busy developing social robots for people with special needs, such as seniors with dementia, children with autism, and adults who have suffered strokes or other such conditions. For them, a Japanese robot resembling a baby seal that responds to stroking and petting, and can distinguish voices is now available. Some critics are wary of such robots, fearing that they will lead to using machines as substitutes for live pets. However, these robots can produce calming and socializing influence on persons with cognitive problems that cause them to feel anxious or isolated. Whereas, live animals require care, are not always available, can be messy, and may pose safety issues, these robots are self-sustaining.

During a trial period in a nursing home it was found that pet robots helped some residents to focus and stay engaged, when their dementia would otherwise make them anxious or wander aimlessly. It was also found that for some of the residents there was lesser need for anti-anxiety medications. While robotic pets are not a complete substitute for human interaction, they could play a vital role in reducing that need for the growing elderly population.

E4 Internet of Things

Today, information technology is largely about data originated by human beings. Now, machines with embedded microchips are able to communicate to each other and with live persons via Internet or other wireless media. This is commonly called the Internet of Things (IoT), which is opening up a broader horizon in information gathering and intelligent control.

In the summer of 2011, a 150-year-old oak tree in a botanical garden in Erlangen, Germany, became an active member of Facebook, Flickr, and YouTube and started to talk. The management of the botanical garden, with the help from a local university, equipped it with sensors and Wi-Fi to collect data. They were doing research on topics, such as how and when the trees actually grow and the difference between the trees in a city with those in a forest. The research that ran for about a year is an example of IoT application.

There is of course much more to the Internet of Things than an interactive oak tree. In the near future motor vehicles will be able to communicate with one another and with their environment to avoid collisions and to optimize traffic flows. Domestic energy users like washing machines, space heaters, and air conditioners will be able to communicate with the electrical network to balance out supply and demand. Machines will thus be talking with other machines like human beings do, via Internet or other wire or wireless media.

Common applications of IoT that are now available include:

- Preventing sudden infant death syndrome: Monitors that provide parents with real-time information about their baby's breathing, skin temperature, body position, and activity level on their smart phones.
- Help in taking medication: A medicine bottle cap with microchip can remind a patient to take medication, inform doctor when the patient defaults, and order the pharmacy to refill when needed.
- Track one's activity level: Sensors, such as blood pressure and heart rate monitor, along with accelerometer, gyro, compass, and GPS can monitor a person's bodily functions on a smart phone and alert a primary care physician if needed.
- Monitor an aging family member: Using wearable alarm button and other wireless sensors placed around the home, a system can monitor the normal schedule and any serious disruption or injury of an elderly person living alone.
- Maintain and repair mechanical equipments including automotives: Sensors installed inside equipment can monitor its mechanical health,

when its parts require service or replacement, and send reports to owners and service providers as needed.

- Monitor atmospheric pollution levels: Multiple air quality sensing systems distributed around a town or city can collect atmospheric pollutant concentrations, thus creating a network of readings for the community and general public.
- Help protect wildlife: Collars that use GPS to locate and track animals.

By 2020, it is estimated that there will be 50 billion devices connected to each other. By then, the IoT will be in use in almost every aspect of our daily lives, from our homes, cars and offices to hospitals, shops, and factories. The Internet of Things revolution will dramatically alter manufacturing, energy, agriculture, transportation, and other industrial sectors of the economy. It will also fundamentally transform how people will work through new interactions between humans and machines.

E4.1 Opportunities and Benefits

Social and technological disruptions will take place because of the availability of massive volumes of data from connected objects, and the increased ability to make automated decisions and take actions in real time. Increased collaboration between humans and machines will result in a much higher level of productivity and more engaging work experience. That will lead to a great increase in the operational efficiency of manufacturing industries and vast improvements in diagnostics and maintenance of consumer products like cars and home appliances.

As the Industrial Internet gains broader adoption, businesses will shift from products to result-based services, where businesses will compete on their ability to deliver measurable results to customers. They may range from guaranteed machine uptimes on factory floors, actual amounts of energy savings, to guaranteed crop yields from an agricultural farm.

IoT will provide new opportunities for people to upgrade skills and take on new types of jobs. The pervasive use of smart sensors, intelligent assistants, and robots will transform the skills mix and focus of the workforce. It will change the basis of competition, redraw industry boundaries, and create a new wave of disruptive companies, like the current Internet, which gave rise to Amazon, Facebook, and Google. Many manufacturing industries are getting interested in IoT, which could lead to a revolution in manufacturing process. For example, raw materials and production machines would be able to enter into a dialogue in order to optimize manufacturing processes with little human intervention.

One can foresees a scenario where distributed intelligence coupled with the ease of communication will create systems that would obviate the need for centralized control and management. That will vastly improve flexibility and manufacturing efficiency making custom single-piece manufacturing economically viable. Internet of Things has the potential to change the world to a greater extent than what Internet has done in the last few decades.

E4.2 Risks and Challenges

The biggest challenges we will be facing are in the areas of data vulnerability and interoperability. Technologists, businesses, and governments will have to work together to overcome these problems. They will also need to address the crucial barrier of interoperability among existing systems. Wider use of Internet of Things will also require seamless data sharing between machines and other physical systems from different manufacturers.

E5 fMRI: A New Tool for Brain Feedback

Emily is a teenager whose mother suffers from chronic depression. Her behavior seemed normal like other children of her age, but their family physician thought that she might be at a higher risk of developing similar depressive symptoms. Therefore, he referred Emily to a medical research facility that specializes in brain disorders. On her first visit, she was evaluated by fMRI (Functional Magnetic Resonance Imaging) of her brain while she was exposed to disturbing images such as car accidents and other catastrophic events (Fig. E2).

During a scan, the investigator monitored blood flow to many parts of her brain, paying particular attention to the amygdale region, which showed greatest activity during stress and anxieties. Emily's blood flow in those parts of her brain was found to be much higher than that of a normal person, particularly when those disturbing images were shown to her. This led to the assumption that she is more at a risk of suffering anxiety and depression disorders, possibly later.

In subsequent sessions, she was shown the levels of her brain activity in particularly sensitive brain regions while being exposed to disturbing images. The investigator then asked her to try to dampen the response by

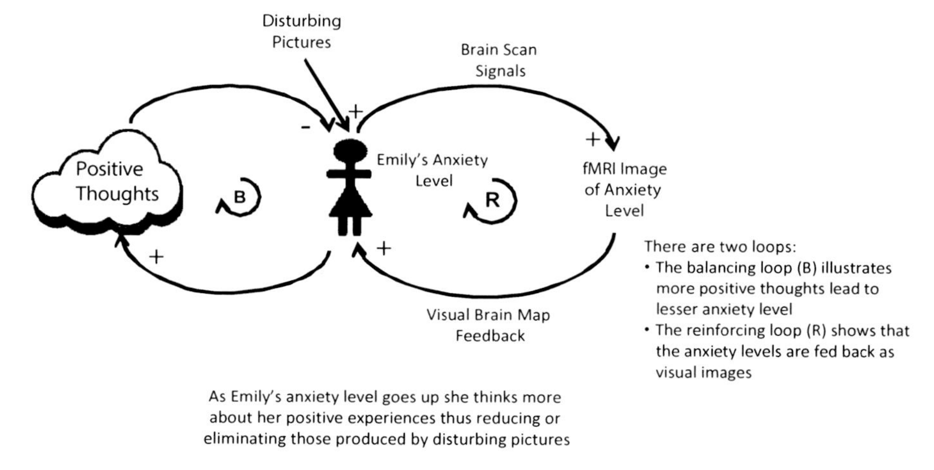

Fig. E2 Brain feedback training to reduce anxiety

thinking about more positive experiences in her own life, such as playing a piano, which she liked, or cuddling her pet dog. To her surprise, she could reduce her elevated brain activity with such positive thoughts. The investigator then asked her to cultivate such positive thoughts, whenever she felt anxious in her daily life and to come back for further fMRI sessions. After a few of these sessions, she learned to dampen her stress reactions sufficiently as her fMRI patterns were reduced to the normal range when exposed to these disturbing images. Emily's case is an excellent example of feedback mechanism where stressful situations may be alleviated by conditioning one's brain response.

E5.1 What Is fMRI?

MRI (Magnetic Resonance Imaging) techniques, which use strong magnetic fields and pulses of radio wave energy, have been used for many years to make static pictures of organs and structures inside a body. Thus, a MRI machine is well suited for imaging brain for injuries, tumors, bleeding, and for other problems, such as damages caused by stroke.

fMRI, on the other hand, is an extension of this technique that indirectly measures the activity of brain cells by detecting changes in blood flow. When an area in the brain gets active, blood flow increases in that area as it consumes more oxygen. Thus, fMRI generates maps showing those parts of a brain that are activated in a particular cognitive or mental process. It is now possible to produce images of blood flow in brains in real time, which has led to the buzzword "real-time functional magnetic resonance imaging (rtfMRI)."

Unlike other biofeedback methods, such as EEG, rtfMRI shows specific areas of activity in the brain due to a particular type of mental stimulation. This precise feedback allows psychologically challenged individuals to learn and self-regulate their thoughts and emotions more easily. That often leads to positive improvements in their cognitive states, which may allow them to reduce or eliminate the need for prescription medications. Emerging evidence also suggests rtfMRI has clinical utility in reducing the symptoms of various ailments, such as chronic pain, tinnitus, and Parkinson's disease. Further trials are going on to draw definitive conclusions regarding the clinical utility of neuro-feedback procedures in a number of such disorders.

E5.2 Comprehending the Physical Basis of Mind

Human beings have been trying for centuries to comprehend the physical basis of what goes on in our mind, in other words, trying to understand the mind–body relationship. With the help of fMRI technology, we can now ascribe the activation of different parts of a brain with changes in the thought process of a subject. Thus, our conceptual barriers between mind and body are starting to breakdown.

As a brain mapping technique fMRI has several significant advantages, such as it is non-invasive, safe, and easy to use, and usually produces images with good resolution. Thus, it is becoming a popular tool for imaging brain function. Over the last decade, it has provided new insight to the investigation of how memories are formed, how languages are acquired, how we feel pain, and many other cerebral processes.

Compared to earlier methods for measuring brain functions, which provided only general view of a brain's behavior, fMRI offers detailed views of the activation of different regions of a brain depending on the changes in the thought process of a subject. This augments our understanding of the localized processes that takes place within our brain for different emotions, thoughts, and actions.

However, it cannot detect the firing of individual neurons or the transmission of information from one neuron to another. Even with this shortcoming, fMRI can still accomplish a lot by being able to pinpoint small regions of a brain that are related to a particular thought process. fMRI often shows activation in more than one non-adjacent region in a brain, showing parallel operations that are taking place for a particular thought process.

Earlier, we discussed how the live feedback from brain could be helpful in reducing the anxieties and other phobias. Other areas where it can be useful include:

Explicit control over brain functions: From our birth, we learn to control our brain functions implicitly. Our brains work implicitly for every action we take, like reading, driving a car, or playing tennis. Here we are unaware of what is going on in our brains and are not actively trying to manipulate them. While meditation and other such practices can produce remarkable results in regulating brain activities, they still are implicit in nature and generally require long periods of training.

Explicit control over the brain, on the other hand, can be learned by looking at brain images and then consciously changing the thought process to fit a particular pattern to produce the desired mental state. We do not yet fully know as to what extent a person can learn to gain greater control over their brain function through brain neuro-feedback, and what will be the consequences of such an exercise. However, fMRI has the potential to bring normally non-conscious brain processes into conscious awareness, and to transform from implicit to a more explicit control as and when needed.

Lie detection: fMRI brain scan may become useful in reading states of people who are unwilling to provide truthful replies to questions. Results of carefully controlled experiments suggest that it is possible to distinguish brain patterns for deliberate dishonesty from that for honesty. This will have dramatic implications, if proven reliable beyond doubt, for screening persons for security clearance or for interrogating crime suspects resulting in improved law enforcement.

Communicating with a patient in a vegetative stage: fMRI can be used to determine the state of consciousness of a patient in a persistent vegetative state due to an accident, a stroke, or some other reason. The patient may be asked to think about something that he or she has experienced in a normal state, like playing a musical instrument or driving a car. The patient would not be considered brain dead if his or her brain scan showed response in appropriate areas. That response could be of immense value, and could potentially lead to life-or-death decisions. However, further research in this area is needed to determine the veracity of such results.

Controlling external devices: Various technologies are being explored to facilitate brain computer interfaces for prosthetic control and other such applications, including EEG and implanted electrodes. fMRI images may provide important insights into how to extend the current limits of what information can be extracted from the brain to manipulate such devices.

fMRI machines are quite bulky and expensive because they use very high-powered magnets that have to be kept at very low temperatures. Researchers are now testing laser-based MRI techniques and other ways of producing three dimensional brain images, which may lead to the development of less expensive and compact devices with substantially enhanced sensitivity. That will give rise to many other interesting applications based on brain feedback.

E5.3 Brain Imaging as an Aid to Meditation

Meditation practices have a long history, especially in Asian countries. In recent years, physicians and others in the western world have employed meditation for successfully treating disorders like depression, anxiety, addictions, and chronic pain. Studies have found that participating in a meditation-training program can have measurable effects on brain functions even when the person is not actively meditating. Thus, meditation practices often produce long lasting benign effects on its practitioners.

Though meditative practices regulate brain activities, their actual effects on the brain itself were largely unknown until recently. EEG sensors that have been in use for some time for measuring brain activities can take readings for about 256 positions around the head thus providing detailed account of parts of a brain that are active during different mental activities like meditating, reading, thinking, and playing. EEG is useful because it is physically non-invasive and does not significantly interfere with meditative practices. However, even with such a large number of sensors, EEG still lacks the precision necessary to trace the complex interaction of neurotransmitter systems. It produces graphical representation of brain waves that require interpretation by experts. Whereas, fMRI images are more comprehensive as they can pictorially display, in three dimensions, the specific areas of a brain that are active. For now, a combination of fMRI and EEG may be used for studying brain activities in research and hospital settings but fMRI machines cannot readily be used in a group meditation session, as they are bulky and expensive.

As new technologies on brain imaging develop, cost effective and easy to use imaging techniques will become available. It would then be possible to use real-time fMRI or other newly developed techniques to give immediate feedback and guidance to meditators and their teachers. That will augment the training and evaluation of meditation practices. For novice mediators it will be a great boon as they may be able to get into deeper meditative states more easily by replicating the brain functions of advanced practitioners. Thus, these new brain feedback technologies offer exciting potentials for future meditation practices.

E6 Gaining System Expertise in an Increasingly Complex World

Rapid advancement of technology, while a boon to human society, is also resulting in societal changes that are leading to increased complexities. Information technology is penetrating every aspects human life leading to the networking of humans, computers, telephones, and household appliances making our life more complex. With the development of Internet of Things, machines will be able to communicate to each other. Soon there will be driverless cars that will be able to talk to each other to inform road conditions, such as fog or icing and warn on traffic jams. As stated before, a driverless car is a very complex system to design and build which will be of significant benefit to many but will also lead to a set of complex administrative, legal, and social problems.

Increased international trade is leading to economic coupling between nations; a recession in China, for example, may lead to economic downturn in the USA and in many other countries. Quicker and easier transportation facilitates the spread of disease more rapidly, thus increasing the chances of worldwide pandemics.

In this highly interrelated world, the political and religious problems in one region are causing terrorism and refugee problems in others. Today, free Internet communication is an essential part of our society, but that is also the leading vehicle for spreading fundamentalism and religious intolerance. Thus, all these technological progresses that are making our lives easier are also making our life more complex and difficult.

E6.1 System Oriented Thinking to Cope with Complexities

A complex system is composed of many interacting subsystems with feedback loops, often with long time delays. A system's behavior is shaped by the information exchanges with other systems and between its own subsystems. However, the web like connections between the components of a complex system often gives rise to unexpected behaviors that are not always foreseen.

System oriented thinking accepts complexity. It acknowledges that one's understanding may be incomplete in many situations, there may be multiple views about a system, and there may be more than one way to solve a problem. Thus, it encourages teamwork and consensus building to allow the team to choose a path that has a higher chance of success.

Synthesis is the most important part of systems oriented thinking. It makes complexity more manageable by encouraging one to step back and see the big picture, rather than focusing on just its parts. It is an effort to see the "forest" as well as the "trees." It explores the interdependencies among the elements of a system, looking for patterns rather than isolated functions or events. It focuses on feedback loops and interactions to understand a system's longer-term behaviors.

E6.2 Teaching System Thinking

Curricula in most primary and secondary schools today are modeled to meet the needs of an industrial society that existed in the era of mass production and assembly lines. At that time, societies and their production methods were stable, where a young person could look forward to a steady job with defined tasks. Then the knowledge and experience gained in the earlier years continued to be useful for the rest of that person's career.

Today we live in a rapidly changing technical and economic environment, where knowledge gained in an academic environment or expertise gained in the first job may become largely obsolete in 5 years. In this changing environment, a student in a school or an undergraduate in a college requires a different type of education.

A young person today needs a deeper understanding of the changing social, technological, political, economic, and other conditions. Solving technical problems may need deep knowledge in chosen fields but the interrelation of various technologies that helps in creating a new product requires a broader view. Prime examples are the development of driverless vehicles, robots to help the elderly and infirm, and enhanced communications between man and machine.

However, technological developments are also leading to wider social changes. Today, there is a crying need to address problems, such as global warming, large-scale unemployment due to automation, or the rise of religious extremism. Young people, today, need to embark on a lifelong education program where the knowledge they have acquired in academic institutions and the expertise they gained in their first job need to be supplemented and often supplanted by new knowledge and expertise.

As stated earlier, the broader view that system orientation brings about is of great value to any person that encounters such a change. Additionally, system knowledge and model building expertise go a long way in increasing the understanding and in addressing these problems.

E6.3 System Orientation as a Part of School Curriculum

System oriented thinking should start from childhood. School curricula need to be updated to make pupils understand the basic principles of feedback. Students should be taught to identify the common behavior patterns of systems and to gain broader perspectives of the problems and their solutions. Other system thinking skills that may be taught include:

- Dynamic nature of systems, where their behavior may change with time or with changes in their environments
- Ability to create diagrams that show the interactions between different components of a system for their better understanding
- Ability to work as a team to solve a problem
- Willingness to learn and accept the different points of views of others
- Ability to understand the difference between short-term and longer-term solutions.

Though system education may start in early childhood, it may not be taught as a separate discipline at that stage. It should be melded with the teaching of other subjects, to extend and energize the teaching of those disciplines. However, that would also require the teachers to understand and appreciate the core principles of system dynamics.

Currently there are a few schools in the USA and in other countries that have included system dynamics in their curriculum and results are largely positive. That needs to be widened to include more students in primary and secondary levels of education. "Creative Learning Exchange," a non-profit organization in the USA is spearheading the teaching of system thinking in schools around the world.

E6.4 Systems Science in Colleges and for Adults

The education on engineered systems has progressed significantly in the last half a century. Many colleges impart basic knowledge of feedback control to undergraduates in engineering and other technical subjects. In addition, a number colleges and universities now offer undergraduate and postgraduate courses on system dynamics as an independent discipline.

However, general public, political, and social leaders do not yet understand that there are deep flaws in the way current policies are formulated. They are mostly driven by short-term pressures rather than by long-term benefits. They remain dependent on mental models and do not understand that human mind is not quite suited for solving higher order dynamic feedback problems. With sufficient systems knowledge they will be able to understand the root causes of these problems and will be in a position to formulate policies that have longer-term benefits. For them there are part-time and web-based courses on system dynamics and complexity. A number of books on system dynamics are also now available for those who want to extend their knowledge in this field.

It is becoming increasingly apparent that system oriented thinking and problem-solving methods are essential in order to survive and prosper in today's complex and rapidly changing world. Today, system expertise is necessary not only for those that are in technological fields but it is also needed for sociologists, economists, politicians, and others.

Appendix I: The PID Controller

A common type of controller used for industrial systems is called a PID or a three-term controller. Where, P stands for proportional, I for integral, and D for derivative control action.

Controller Actions

Proportional Control Action Proportional control is the most common and intuitive type of control action, where a controller's output is proportionally related to the error that needs to be corrected. The term "error" signifies the difference between the set point or desired value and the actual value of a variable. As is the case with the Watt Governor, where the steam flow to an engine is proportionally increased or decreased when it slows down or runs too fast.

A proportional control action deals only with the present condition in a system. It does not take into consideration the past history or possible future consequences. The main limitation of proportional control is that it cannot maintain the process output exactly at the set point. That is because the proportional action produces a corrective output only when there is an error. In other words, there is no controller action if there is no error, and smaller the error weaker is the corrective output. In a steady operating condition, the difference between set point and the actual value is called offset (Fig. I.1). Increasing the gain, which is the ratio between the changes in output to a change in input, reduces this offset, but too much gain makes a system unstable. That leads to oscillation, which can only be corrected by reducing the gain.

A. Ghosh, *Dynamic Systems for Everyone*,
DOI 10.1007/978-3-319-43943-3

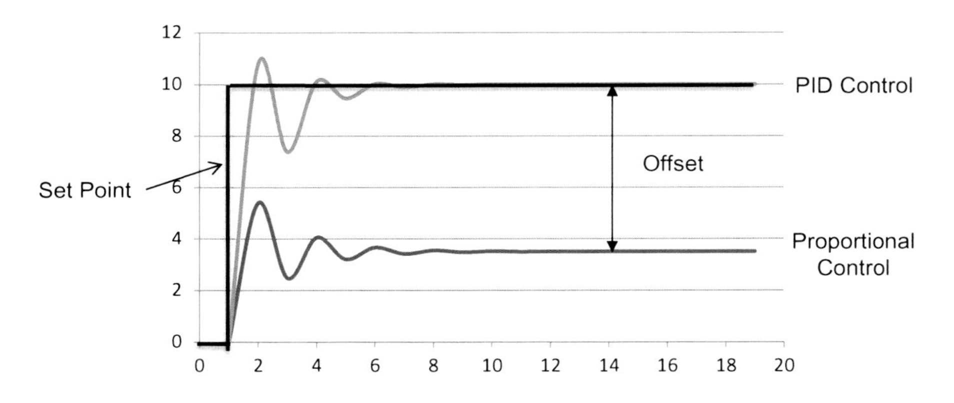

Fig. I.1 Proportional only and PID control

Integral Control Action Adding integral control action eliminates the offset error without decreasing the system stability. The output of integral action is the product of integral control gain and the sum of errors in the immediate past. Thus, when there is even a small error, its sum increases with time leading to an increased change in the controller output, which reduces and then eliminates the offset.

Derivative Control Action While proportional control action considers the present error and integral action looks at the cumulative sum of errors in the immediate past, the derivative action looks at the rate of change of error. Thus, the derivative control somewhat anticipates and acts on the future value of the error. In an industrial application, derivative control reduces offset and the period of oscillation faster than a system with only proportional and integral control. However, derivative control amplifies the noise in a system, so it is generally not favored when such a condition exists. In most applications, the derivative control action is kept at a very low setting to minimize the effect of noise or sudden changes in set point or the measured value.

Proportional (P), integral (I), and derivative (D) are the three most commonly used types of control in industrial applications; each has its own advantage and disadvantage. They are seldom used alone, when control requirements are rigorous because of their inherent limitations. Most modern industrial controllers (known as PID controllers) incorporate all the three types of control actions. There, the outputs of integral and derivative control actions are added to (or subtracted from) the proportional control action output to provide a single signal for corrective action.

Loop Tuning

A set point change or a disturbance in a system causes a process output to oscillate before it settles down. In most applications, this oscillation is hard to eliminate totally. Reducing the gain reduces the oscillation, while increasing the gain amplifies it. However, if the gain is reduced significantly, then the system gets sluggish and the controlled variable takes significantly longer to reach the desired value. In most cases, a compromise is necessary whereby a system does not oscillate too much or get very sluggish. A quarter (1/4) decay ratio is commonly acceptable, where the amplitude of each oscillation is one quarter of that for the previous oscillation peak (Fig. I.2), that requires proper tuning (adjustment) of the controller's proportion, integral, and derivative actions. As there are enormous variations in the characteristics between different processes, each control loop needs to be tuned individually for their optimal performance.

Ziegler–Nichols Method There are various rigorous and ad hoc methods for optimum tuning. Among them, the Ziegler–Nichols method is the one that is the simplest and possibly the most commonly used.

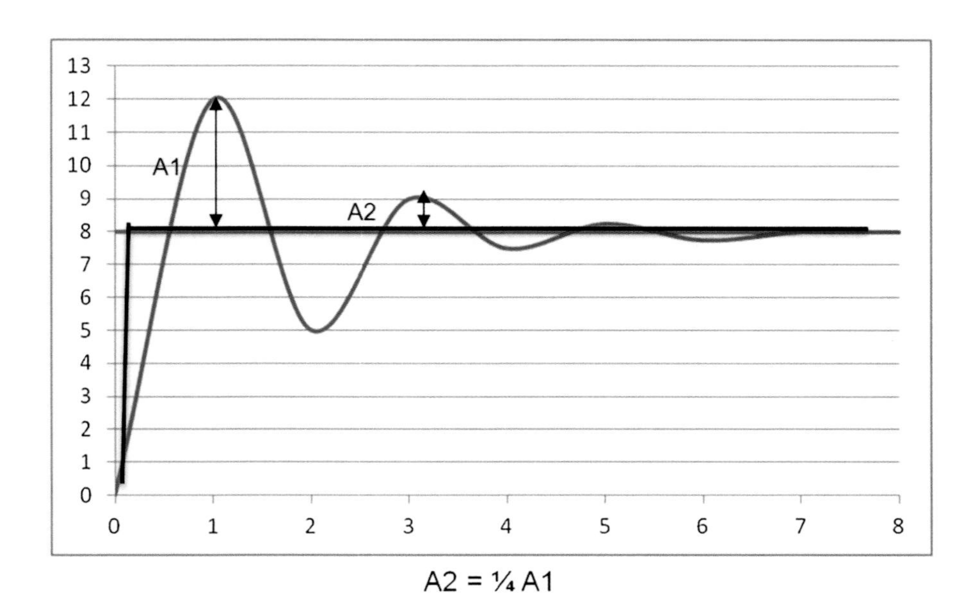

Fig. I.2 Loop tuned for quarter decay

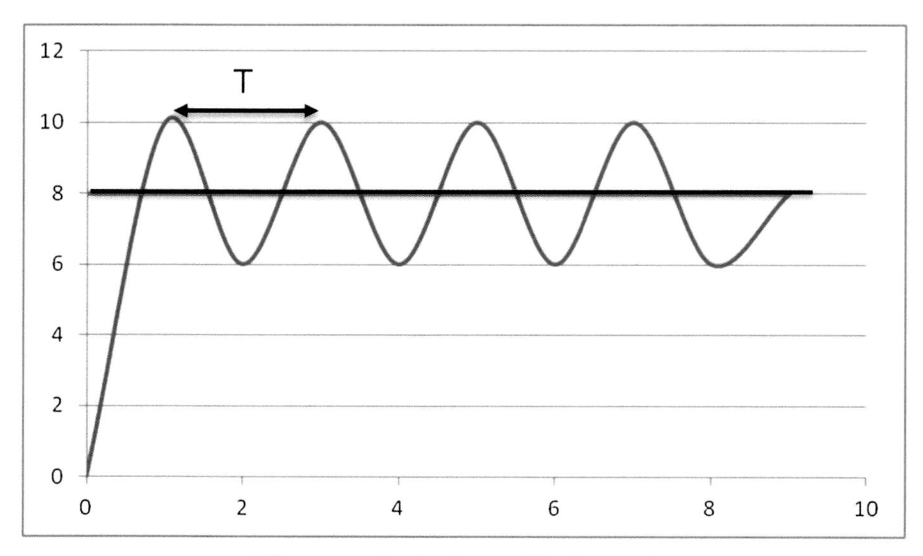

Fig. I.3 Oscillation with constant amplitude

A step-by-step approach to Ziegler–Nichols method of tuning a controller:

- Disable integral and derivative actions and increase the proportional gain until the output variable to be controlled starts a sustained oscillation at constant amplitude (Fig. I.3). Note the gain K and the period of oscillation T.
- If using proportional control only, then set the gain to half (0.5 K).
- If using both proportional and integral control, then set proportional gain to 0.45 K and integral gain to 0.55 K/T.
- If using proportional, integral, and derivative control, then set proportional control to 0.6K, integral gain to 1.2K/T, and derivative gain to 0.08 KT.

This is a good starting point to observe whether a system so tuned is adequate for the application, which is true for most applications. Further adjustments may be made where necessary, such as increasing or decreasing proportional gain to get the desired behavior.

For a system that has a large number of control loops, proper tuning of all the loops is laborious and time-consuming. The good news is that software packages are now available for tuning the loops automatically, thus reducing human intervention to a minimum.

PID Control Equation

PID control may be represented by a mathematical equation, which specifies the output of a PID controller based on an input error. There are various versions of this equation, and here is the one.

$$C_o = K_p e + K_i \int e dt + K_d de / dt + B.$$

$K_p e$ = Proportional control action
$K_i \int e dt$ = Integral control action
$K_d de/dt$ = Derivative control action

Where:

C_o = Controller output
e = Error
K_p = Proportional gain
K_i = Integral gain
K_d = Derivative gain
B = Bias

Error: Difference between set point or desired value and the actual value
Gain: The ratio between the changes in output to a change in input. Thus, higher the gain, higher is the output for a given input
Proportional gain: Gain that is proportional to error
Integral gain: Gain that is proportional to the sum of immediate past errors
Derivative gain: Gain that is proportional to the rate of change of error
Bias: Output of a controller when there is no error input; this is usually set manually to reduce the offset.

Appendix II: Guidelines for Drawing Causal-Loop Diagrams

English narratives can adequately convey the state of a system at an instant or its progression over a period of time, but is not very helpful in showing the cyclical interrelations between cause and its effects. Here, a causal-loop diagram helps us to articulate and to enhance our understanding of a dynamic system.

Before generating a causal-loop diagram, it is important to decide on its purpose and its scope. Without that, a diagram can end up as a jumbled set of variables and arrows. It is important to consider the cause and its effects within its normal operating range when their behavior is predictable. It is also helpful to decide on the time horizon of the behavior of the variables that are to be depicted.

How detailed and at what level should a diagram be? That depends on the issues that are under consideration. For the ease of understanding, a set of minor cause and effects may be lumped together as a single variable, which may be elaborated in another causal-loop diagram. To make a diagram that is easy to generate and understand, it should be reproducible on a single sheet of a standard-sized paper.

Some Simple Rules Here are some general rules for generating causal-loop diagrams (Fig. II.1). This should be read in conjunction with the relevant sections in Chap. 3.

1. Use a noun or noun clause to specify a variable and avoid phrases that specify action. For example, use a phrase like "Sales volume" rather than "Increase in sales" or "State of the economy" rather than "Growing

A. Ghosh, *Dynamic Systems for Everyone*,
DOI 10.1007/978-3-319-43943-3

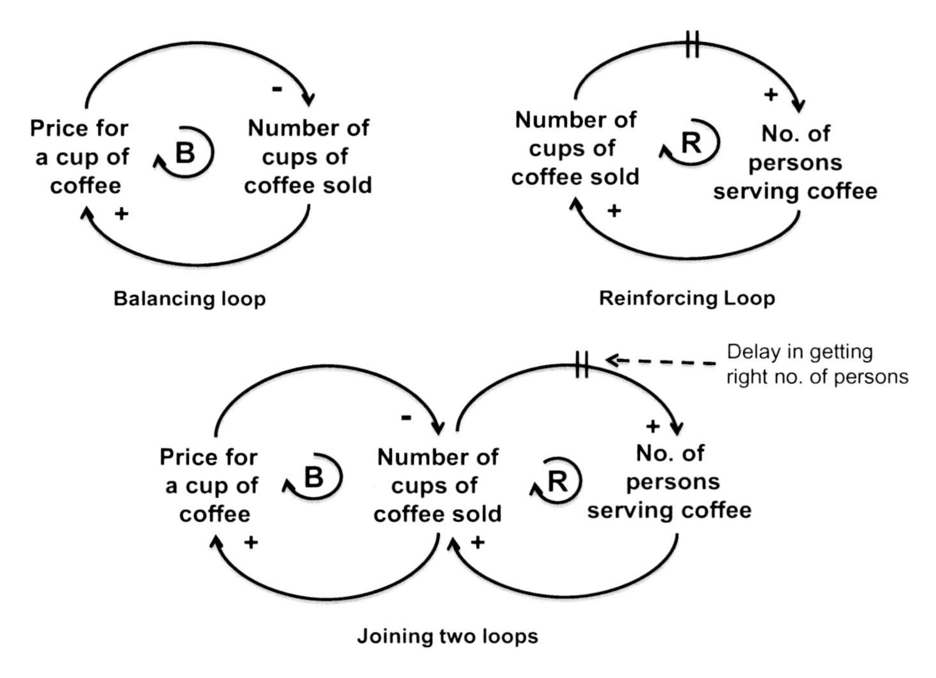

Fig. II.1 Drawing causal-loop diagrams

economy." "Population growth rate" is appropriate, but not "Increasing population."

2. Specify those variables that are changing or will be changing during the period under consideration. It makes little sense to specify things that are static.
3. Specifying a variable name in a positive sense makes it easier to understand than by using negative sense. Using "Growth rate," for example, rather than "Rate of decline." Then a contraction may be expressed as negative growth rate.
4. Time itself should not be depicted as a variable. It is often implicitly specified by an arrow that connects between a cause and an effect. Nevertheless, large lags or delays should be specified explicitly by inserting a double hyphen (=) on an arrow, that connects the cause and its effects.
5. When a cause and its effect move in the same direction, then the arrow connecting them should show a positive sign (+) at its head, while a negative sign (–) should indicate that they move in opposite directions.
6. All major feedback signals should be specified along with possible consequences. They may be lumped together where necessary to improve clarity.

7. A loop with no negative sign or an even number of negative signs is a reinforcing loop, which should be indicated by an "R" with a round arrow showing the direction of influence. A loop with an odd number of negative signs is a balancing loop, which needs to be indicated by a "B" with a round arrow indicating the direction of influence.
8. Two loops may be joined together when they are using a common variable. That generally improves the understanding of a system's behavior. However, that may be avoided, if a diagram gets too large or complex.

Appendix III: Generic System Behaviors

There are a number of common behavior patterns in social, economic, political, and other types of systems. Donella Meadows was one of the first to recognize them as system archetypes. They are generic in nature and are not meant to represent complete systems. They occur repeatedly in various systems or as parts of systems. It is helpful to understand them, so that they may easily be identified when a system exhibits problematic behavior.

Limits to Growth

We all know that in real life nothing can grow forever. A growing action is always balanced by a slowing action that limits the unfettered growth of a variable. Finally, a limiting condition may stop the growth altogether. The causal-loop diagram shows a reinforcing loop for the growing action coupled with a balancing loop for the slowing action along with the limiting condition (Fig. III.1).

A child cannot physically grow forever. First, the growth slows down, and then the child reaches a limit during late teenage years due to the natural regulation of his or her growth hormones. Similarly, a fast spreading pandemic eventually slows down, either because of the demise of most of those that are vulnerable, or due to the preventive measures taken to stop its spread. The fast sales growth of a novel product made by a manufacturing company may soon encounter slower growth due to various internal factors, such as dearth of trained personnel or capital to increase its production capacity indefinitely. Externally, it may encounter increased competition from other companies

A. Ghosh, *Dynamic Systems for Everyone*,
DOI 10.1007/978-3-319-43943-3

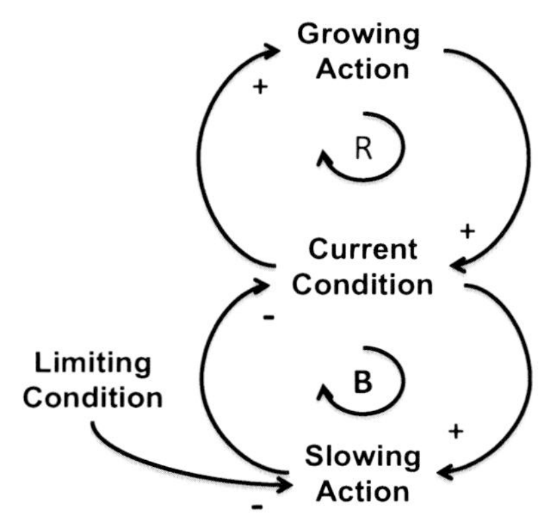

Fig. III.1 Limits to growth

that start making similar or better products. If none of these happens, then market saturation will certainly put a hard limit to its growth in a not too distant future. The obvious solution for a problem of this type would be to weaken or eliminate the slowing action loop, if possible.

Shifting the Burden

On encountering a problem, a common tendency is to fix its symptoms rather than addressing its underlying cause. That is by shifting the burden from the fundamental cause to its symptoms. Fixing a symptom is often a low-cost solution, which produces quick result, but that will not generate long-term benefits and may also cause additional complications or side effects. The causal-loop diagram shows a balancing loop for system-based solution and another for solution based on fundamentals, along with a delay. There is also a third reinforcing loop which shows the side effects associated with symptom-based solution (Fig. III.2).

There are many examples of shifting the burden in real life, such as taking aspirin to reduce fever, which quickly reduces the body temperature and brings fast relief, but does not address the underlying cause of the malady. Moreover, a prolonged use of aspirin can produce additional complications. There are many examples of shifting the burden in a contemporary political climate, where politicians try to solve major systemic and economic problems by quick fixes rather than addressing the underlying causes. A symptom-based quick fix is sometimes desirable, but that should not lead us to ignore the underlying cause of a problem.

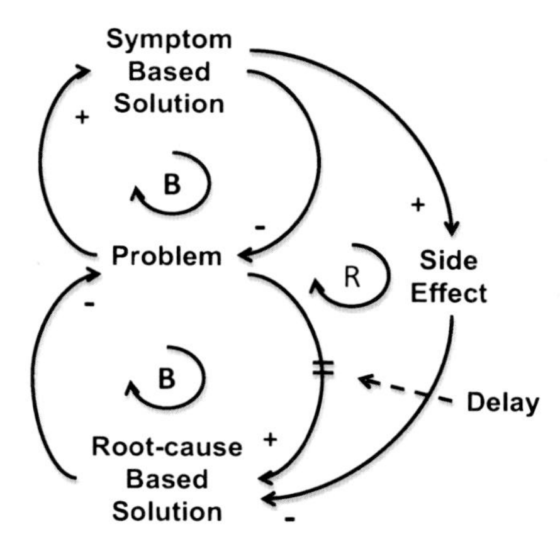

Fig. III.2 Shifting the burden

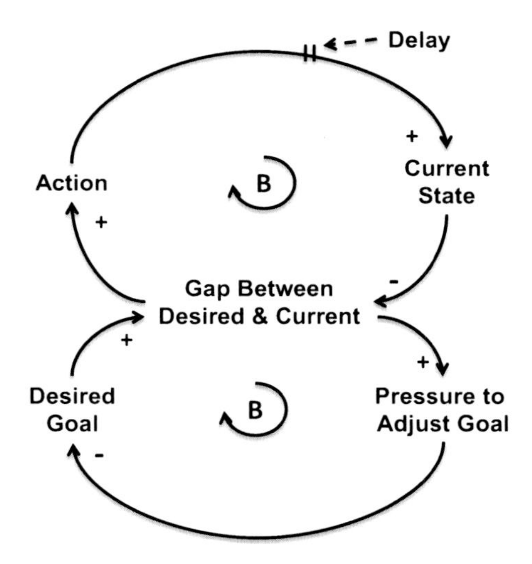

Fig. III.3 Shifting the goal

Drifting Goals

When an individual or an organization sets a goal but subsequently finds it hard to reach, a common tendency is to change the goal to something more achievable. The causal-loop diagram shows one balancing loop that is trying to reduce the gap between the desired and current state by taking appropriate action, while the other balancing loop is trying to reduce the gap by shifting the goal itself (Fig. III.3).

At first, that may seem to be quite a rational thing to do. However, by repeatedly allowing the performance standard of an individual or an organization to be lowered based on past performance sets it to a path of continuous drift towards lower performance. We see examples of that in public life, where the standards of many primary and secondary school systems get lower every year without causing much concern to the community at large until the system is totally broken. The way out for such a situation is to steadfastly cling to the original goal and try to fix the problems that are causing the lower performance in the first place.

Escalation

When nation X increases its arms buildup, nation Y, which has an adversarial relationship, also increases its defense budget to match or better X's perceived military power, that in turn influences nation X to spend even more on its defense, and thus, setting up a vicious circle. The two balancing loops try to reduce the perceived gap of an attribute between the two independent groups of actors, which lead to an escalation (Fig. III.4).

The scenario above was typical in the cold war between the West and Russia and is still the case with many other nations in this world today. Similar situations

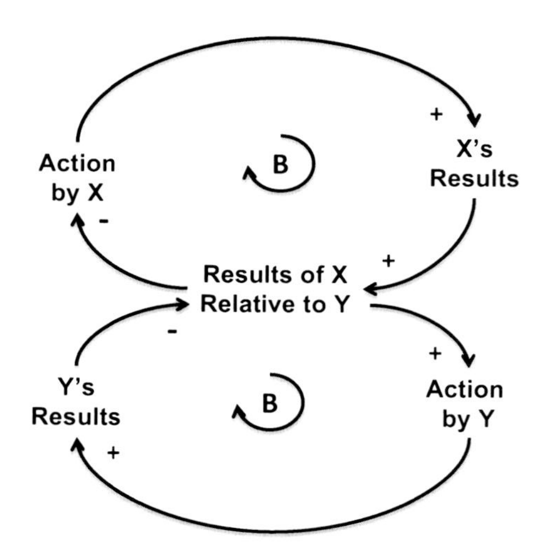

Fig. III.4 Escalation

happen in gang rivalry, where one gang's violence towards another is met with greater violence in return, thus escalating the warfare between groups. Often, the magnitude of these threats are perceived rather than being actual, where prudence and statesmanship are needed, so that the gang violence does not get out of hand or rivalry between the nations is constrained. Cooperation rather than competition can also reduce such an escalating problem.

Escalation has a bright side, for example, free competition between businesses benefits the consumers, which is considered as a foundation of a capitalist society. Escalation also has other positive effects, such as increased civility by a person or a group is often reciprocated by others, thus promoting increasing civility within a society or among nations.

Success to Successful

If a child does academically better than her siblings, then that child usually gets more attention from teachers and parents, while her brothers and sisters get lesser attention and support. That usually leads to the other siblings falling behind even farther. The academic performance gap widens between the bright one and her siblings, even though their relative academic intelligence gap may not be that wide. This is an example of success to successful, which is depicted by two reinforcing loops, one augmenting the resources to a more successful individual or organization, while the other is reducing the resources to those who have been less successful (Fig. III.5).

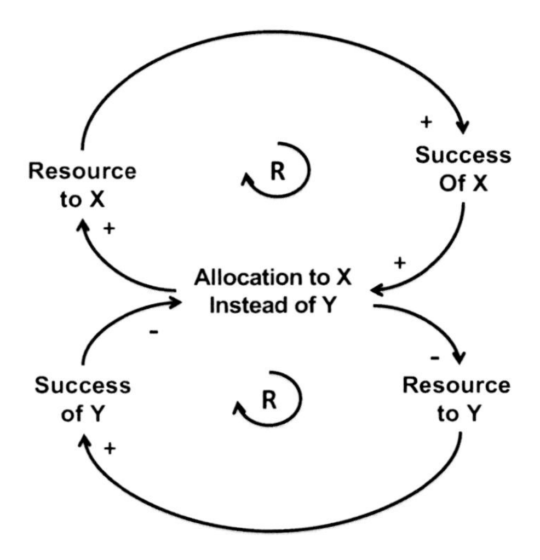

Fig. III.5 Success to successful

Similar things happen in large commercial organizations, where those branches that are initially more successful usually get greater support than those that are lagging behind. That makes successful branches more successful even when the difference in business talents of personnel between these branches may not be that different. Thus, initial conditions often determine the winners and losers.

Even though intelligence, hard work, and business talents ought to be prized, giving unfair advantage to those who had better initial conditions, does not serve well for individuals, businesses, and the society at large. Thus, an effective implementation of antitrust laws and offering more level-playing fields to businesses and individuals can go a long way in creating more just and prosperous societies.

Tragedy of the Commons

In England, a "common" is a piece of land that is available for everybody's use. There cowherds may bring their cattle for grazing without any restriction. The problem with such an arrangement is that a cattle owner would try to maximize his or her own benefits by grazing as many animals, for as long a time as possible, without taking care of the grazing area. That leads to overgrazing, causing deterioration of the soil and vegetation, thus adversely affecting all cattle owners in the long run. The causal-loop diagram shows two reinforcing loops, one for each cattle owner who individually gains more by overgrazing. However, two other balancing loops, which take into account the total activities and limited resources, reduce the gains made by the individual cattle owners in the end (Fig. III.6).

This problem is all too common in modern day life, where we act to maximize our immediate individual benefits, without looking into the longer-term implications of our actions. We as individuals and members of societies, business organizations, and nations use common resources like air, water, fossil fuels, minerals to maximize our individual or organizational benefits without due regards to the finiteness of their quantity or quality. It is only a matter of time before the adverse effects of their overuse become apparent. This is in contrast to the traditional tribal societies that managed common resources more judiciously by avoiding their over-exploitation.

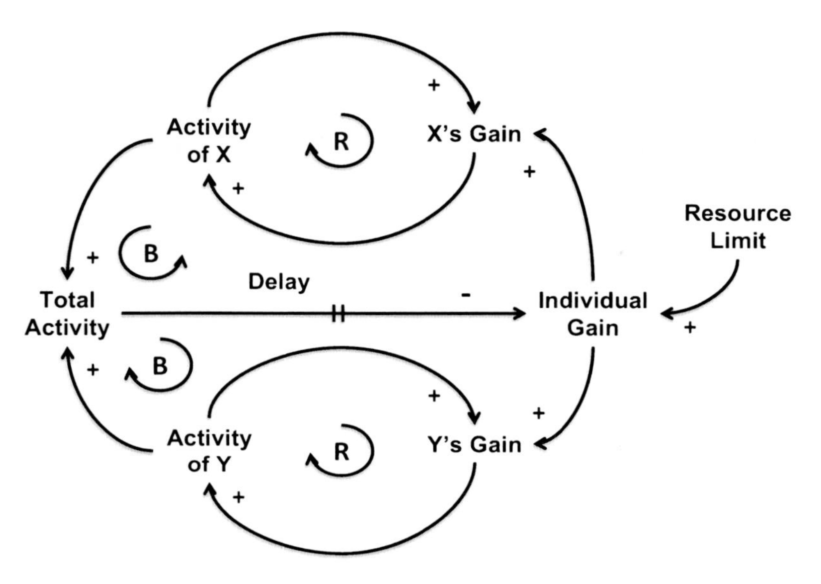

Fig. III.6 Tragedy of the commons

Fixes That Fail

We often find that after a problem is fixed, it reappears later in a similar or in a modified form. The reason may be a byproduct or unintended consequence of the applied fix. The causal-loop diagram shows one balancing loop for a problem with its fix, while a reinforcing loop includes the unintended consequences caused by the fix, which reduces or negates the effects of the fix (Fig. III.7). For example, China had an overpopulation problem, the arable land there was not enough to sustain the natural population growth; at least that was the opinion of the ruling communist party. Therefore, they introduced a one-child policy, which slowed down the population growth. However, male babies are prized in China, which led many families to kill their new born if they were female, causing serious gender imbalance. Additionally, the one child policy is increasing the proportion of senior citizens dramatically with a smaller number of younger people to support them.

The effects of such unintended consequences are discussed in greater detail in Chap. 9.

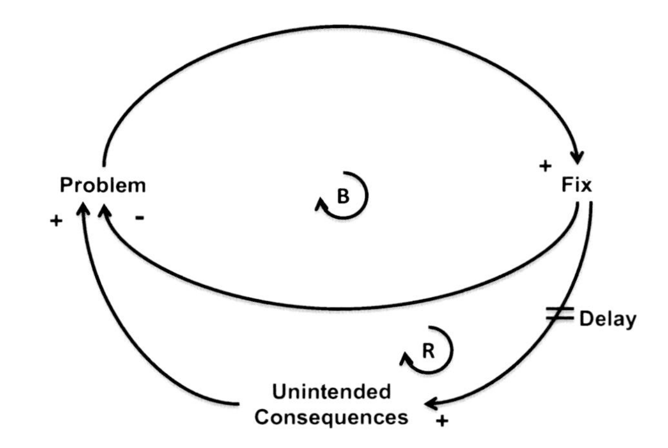

Fig. III.7 Fixes that fail

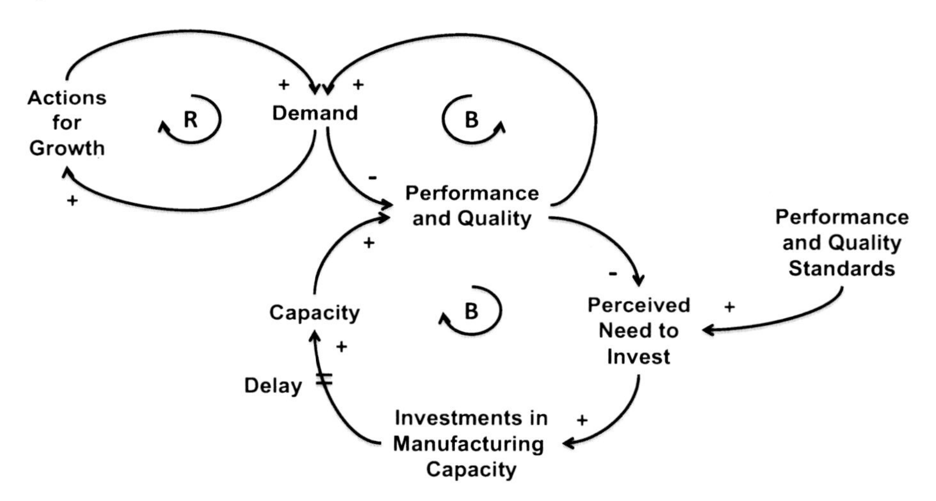

Fig. III.8 Growth and underinvestment

Growth and Underinvestment

A fast growing manufacturing company often flounders, when it fails to invest in all areas of its business. It may decide to increase the number of sales personnel to sustain the sales growth, but may fail to invest enough in production and quality control. That may lead to lower product quality and increased customer dissatisfaction causing decline in sales and, in some cases, eventual downfall of the business. The causal-loop diagram shows three inter-connected loops (Fig. III.8). The loop at the top left on the diagram shows the reinforcing relationship between a growing action and the demand, and the loop at the top right shows the balancing action between demand and performance that indicates that the increase in demand can reduce the performance or quality of a product. The third loop at the bottom shows the balancing action between the performance, the need to invest in capacity, and the increase in capacity.

Similar situation can arise when there is increased investment in manufacturing, but little in after-sales support. Thus, a balanced investment on all aspects of the business activity can avoid such problems.

Appendix IV: Model Building and Simulation Software

Model Building is Fun, but can be Challenging

The Fun Part Building and playing with models has always been a popular hobby for children and for many grownups. In my childhood, I played with an Erector set, which allowed me to build many different toys and structures like trains, boats, airplanes, and bridges from a set of common parts. Today, children play with Lego and get similar enjoyments in building various objects. They experience vicariously some of the thrills and excitements to operate them, which are not readily available in their real life. That explains the popularity of video games amongst the adolescent population around the world. However, the video games are mostly prepackaged, leaving little chance for the players to alter their basic functions.

It is, however, more exciting to both young and old when they are able to build models and observe their expected and unexpected behaviors. The good news is that today, personal computers are powerful enough to run many simple to sophisticated modeling packages. The increased use of personal computers in schools is making the students more computer savvy, thus increasing their interest in simulating real-life problems when they grow up. Computer simulation is also being used increasingly in secondary schools and colleges to impart knowledge in various disciplines ranging from physics and mathematics to biology and social sciences. But what about those of us who are already grown up? Downloadable and easy-to-use modeling packages are now available for free or at nominal price. The instructions for their use are provided with the package or are available at the suppliers' websites.

A. Ghosh, *Dynamic Systems for Everyone*,
DOI 10.1007/978-3-319-43943-3

Today, building a model is not very difficult even for a novice. One may start with a relatively simple system, draw a causal-loop or any other diagram to depict the system interactions on a piece of paper, and go from there. The websites of the suppliers include many prebuilt and tested models. It is easier for a beginner to pick a suitable model from a supplier's websites and then modify it to fit his or her particular interest rather than building it from scratch.

The Challenges The process of generating a model for the first time can be harder than it looks, but one does not need to feel discouraged. It is crucial to define the nature of the problem early in the project. Lack of clarity of purpose for generating a model can lead to a lot frustrations and major reworks later.

It is important to choose a right modeling package for an application. Decide on whether to generate a traditional system-dynamic model or an agent-based model. In most application, a traditional modeling approach will suffice, where the focus is on the behavior of a class or a group, such as the simulation of predators and prey, where a group of wolves and a group of deer are considered. However, an agent-based modeling package should be used if the focus is on the behavior of individual wolves and deer.

One should start with a simple model and then expand it, keeping in mind the ultimate purpose of the exercise. If the model to be built is for a group of people, then active involvement of all the members is essential. As there is a need for shared understanding between all members of the group on the aim and the logistics of building the model.

A model builder needs to get familiar with the software package early in the project. However, it is not necessary to know all the detailed capabilities of the package, which may be learned with the progress of the project.

Software Packages for Model Building and Simulation

There are a number of software packages available for building interactive models that can run on personal computers. They include iThink/Stella, PowerSim, Vensim, Insight Maker, and Simulink. Brief accounts on these packages are given below. Additional information on these packages is available at their websites.

iThink and Stella

isee systems: http://www.iseesystems.com

iThink and Stella are two different versions of the same software package, where iThink is oriented towards business and commercial use and Stella for academic and personal use. They provide graphically oriented front-end for the development of these models. The stock and flow diagrams, used in system-dynamics literature, are directly supported with a series of tools for model development. Dialog boxes accessible from the stock and flow diagrams allow writing equations. They are available for both Macintosh and Windows-based PCs.

PowerSim

Powersim Software AS: http://www.powersim.com

In the mid-1980s, the Norwegian government sponsored research aimed at improving the quality of high school education using system-dynamics models. This project resulted in the development of Mosaic, an object-oriented system aimed primarily at the development of simulation-based games for education. Powersim was developed from Mosaic for generating system-dynamics models for a learning environment.

Vensim

Ventana Systems, Inc.: http://www.vensim.com

Originally developed in the mid-1980s for use in consulting projects, Vensim was made commercially available in 1992. It is an integrated environment for the development and analysis of system-dynamics models. Vensim's user interface is similar to that of iThink and Stella. Vensim runs on Windows and Macintosh computers.

Insight Maker

Insight Maker: http://insightmaker.com/

Insight Maker is a web-based modeling and simulation package. It runs on Insight Maker's website, which can be easily accessed from any user's PC. Once a model is built, it is stored in the same website, which may be shared by any other user who can modify or update the model after making a copy. This helps creating a user community that cross-fertilizes each other's ideas. The best part of Insight Maker is that it is completely free after a user registration at their website. The user interface of Insight Maker is similar to that of iThink and Stella.

Simulink

MathWorks: http://www.mathworks.com/products/simulink/

Simulink provides an interactive graphical environment and a customizable set of blocks that let a user design, simulate, implement, and test a variety of time-varying systems, including communications, controls, signal processing, video processing, and image processing. It is a subset of the MATLAB package offered by MathWorks and it offers close integration with the rest of the MATLAB environment. It is widely used for modeling many mechanical, electrical, and industrial systems. It is also widely used in colleges and universities for teaching engineering and physical sciences.

Spreadsheets for Modeling and Simulation Spreadsheets are widely used for various numerical calculations. Most of us are quite familiar with Microsoft Excel or other spreadsheet software packages. So the question is—can we use them for system modeling? The answer is yes, but simulation of some of the modeling functions are not that straightforward. In a spreadsheet, linear relationships are easy to specify, but that is not the case with nonlinear functions, whereas most simulation packages provide many built-in nonlinear functions. Moreover, most systems have delays, which are not easy to simulate on a spreadsheet. Therefore, common simulation packages are preferred over spreadsheets by a majority of model builders.

Relative Strengths of the Simulation Packages

Stella, Simulink, and many other simulation packages are versatile enough to simulate many different types of systems. However, iThink/Stella, Vensim, and Insight Maker are easier to use than Simulink as they make lesser demand on mathematical knowledge. The basic concepts of Stella were developed by Jay Forrester and his group at Massachusetts Institute of Technology's (MIT's) Sloan School of Management for the simulation of business and social systems. It has since been extensively used to simulate various other types of systems, such as political, economic, biological, and other natural systems. Stella is also used to impart system knowledge to schoolchildren.

Insight Maker, as mentioned above, is a free web-based software package that includes many example models. That makes it an ideal package for the beginners.

Appendix V: Simulation of Manpower Needs for a Project

This is an example of a simulation of work force requirements for a project that is running late. It shows that increasing the number of persons in the project reduces the time for its completion, but increases the total cost because of the additional training requirements and increased overheads. Introductory details and the conclusions of this simulation are in Chap. 6.

The software project Sam is leading is behind schedule. According to his best estimates, it will take over a year to complete the project with the present workforce of ten engineers. This is unacceptable to the client, who wants it to be ready and running in 9 months or less. Therefore, he needs a few additional engineers to complete the project on time.

Sam divides the project up into 1000 tasks of similar size and estimates that each engineer will nominally be able to complete 10 tasks in a month if he or she is able to work in isolation without the burden of communication and coordination with the rest of the team members. However, in reality that is not possible. The team has a flat structure, where the members communicate with each other as needed to make sure that the various bits of software they produce are compatible to each other and their interactions are properly coordinated. He needs to maintain that arrangement.

With his experience on earlier projects, he estimates that the average communication and coordination effort between any two persons in the team takes about 2 % of each of their working hours.

Therefore, by multiplying the number of persons in the team, less one, by 2 %, he would get a reasonably accurate figure of the ratio of the total amount of working time each person will be spending on communication

A. Ghosh, *Dynamic Systems for Everyone*,
DOI 10.1007/978-3-319-43943-3

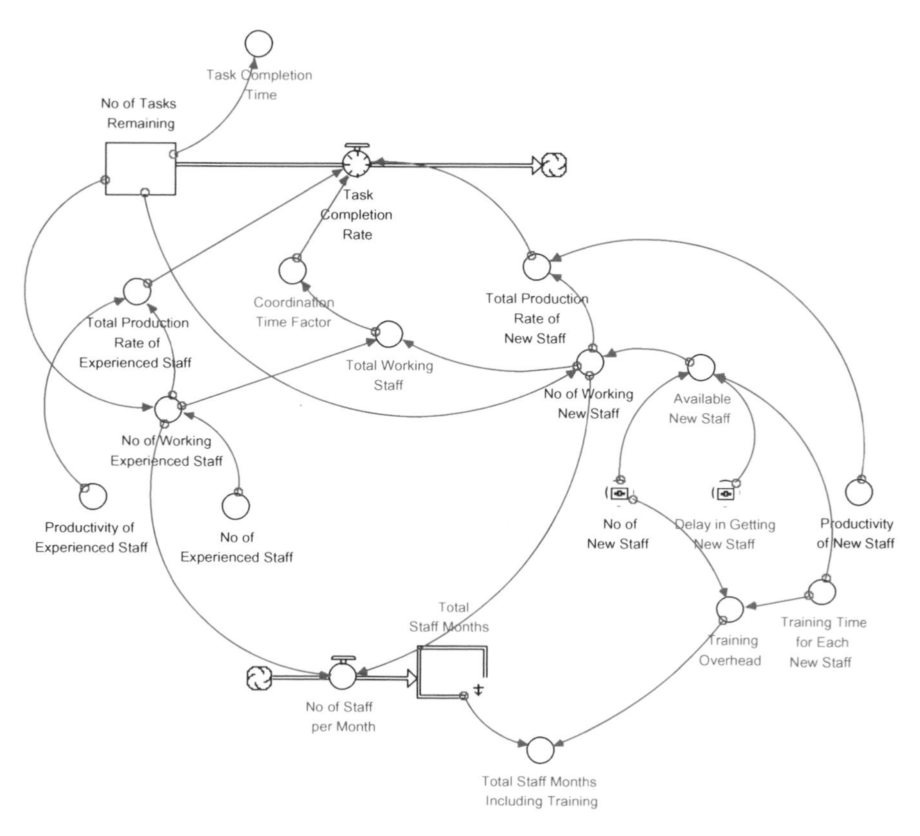

Fig. V.1 Model of project manpower needs with Stella

and coordination, which is (2/100)*($n-1$), where "n" is the number of members in the team.

Sam builds the model, using Stella modeling package, with two stock items—number of tasks remaining for completion and total staff months expended (Fig. V.1). On project completion, the number of tasks remaining depletes to zero, while total staff months reaches its maximum value. The depletion rate (outflow) of the tasks remaining is set by the total productivity of the available staff after training, while taking into account the communication and coordination efforts. The accumulation rate (inflow) of the total staff months expended depends on the total staff employed in the project regardless of whether they are productive or in training.

Once the model was complete, Sam was able to make a number of runs with varying numbers of new staff. The simulation produced figures for project

Table V.1 Simulation of project manpower needs with a varying number of staff

Run no.	No. of New staff	Total staff Months	Percent time for coordination With full team (%)	Project Completion Time (months)
1	0	122.5	18	12.25
2	1	127.0	20	11.50
3	2	130.0	22	10.75
4	3	134.8	24	10.25
5	4	138.8	26	9.75
6	5	145.0	28	9.50
7	6	147.0	30	9.00
8	7	152.3	32	8.75
9	8	157.0	34	8.50
10	9	161.3	36	8.25
11	10	170.0	38	8.25
12	11	173.0	40	8.00
13	12	182.0	42	8.00
14	13	190.5	44	8.00
15	14	193.0	46	7.75
16	15	201.3	48	7.75

Given:
Number of experienced staff: 10
Productivity of experienced staff: 10 tasks/month
Time when new staff will be available: In 1 month
Training period for each new staff 1/2 month
Productivity of new staff after training: 10 tasks/month
Percent time each person spends on coordination with another staff member: 2 %.

completion time, the total staff months, and the percentage of time spent on coordination efforts (Table V.1). He was then able to generate graphs for project completion time and total staff months for various numbers of additional team members, see Chap. 6 (Fig. 6.2).

Appendix VI: Agent-Based Modeling Software

There are a number of agent-based modeling packages that run on Windows and Mac Operating System platform and the list is growing. Many of them are freely available and can readily be downloaded from their websites. Notable among them are

NetLogo: http://ccl.northwestern.edu/netlogo/
Repast: http://repast.sourceforge.net/#
StarLogo: http://education.mit.edu/starlogo/
Swarm: http://www.swarm.org/wiki/Main_Page

Among these, NetLogo is one of the most widely used packages. It is comparatively simple to use for students, teachers, and intelligent novices, but is advanced enough for researchers to generate models in many different areas of study. NetLogo is free, which may be downloaded from its website.

NetLogo

NetLogo was created by Dr. Uri Wilensky, Professor at Northwestern University in Evanston, Illinois, USA. NetLogo is considered to be a dialect of the Logo language, which was designed earlier by Professor Seymour Papert at MIT, who created Logo as a tool to improve a child's thinking, where a small robot called "turtle" is used to solve problems. The word "Net" is meant to convey the decentralized and interconnected nature of the agents, which

A. Ghosh, *Dynamic Systems for Everyone*,
DOI 10.1007/978-3-319-43943-3

are called "turtles." These turtles move around an area called the View, which is made up of a grid of patches.

NetLogo is designed for modeling complex systems with many different agents or actors. A user may simulate a large number (hundreds or thousands) of agents, all operating independently under different conditions. NetLogo is unlike the system-dynamic modeling software packages discussed earlier, such as Stella, Insight Maker, and Simulink, where the cause and effect relationships are between classes or groups, but not between individual actors. NetLogo allows a user to explore the connection between micro-level behavior of individuals and the macro-level patterns that emerge from their interactions.

In NetLogo, turtles may be made to represent as many different types of agents as one can imagine, such as animals, birds, bees, ants, bacteria, people, tribes, voters, citizens, buyers, sellers, metals, cars, trucks, robots, molecules, electrons, and neutrons. Patches can simulate lawns, forests, farmland, trees, houses, rivers, and the like. Turtles and patches can be used to create art, play game, and study mathematical abstractions.

NetLogo website has an extensive set of documentation and tutorials. There is also a library of prebuilt models that can be used and modified. These models cover a wide area, such as physics, chemistry, mathematics, computer science, biology, medicine, economics, and social psychology. A beginner may follow the tutorials that are in the NetLogo website and try modifying one or more models from the library before embarking on building a model from scratch. The main limitation of NetLogo is the limited learning capabilities of its agents.

NetLogo Environment

Traditionally, NetLogo environment is in two dimensions, length and width, like a flat piece of paper. A three dimensional version with length, width, and height, like a box, is now available. Following is a brief description of NetLogo's two-dimensional environment. Further details on both these versions are in the NetLogo website.

NetLogo is made up of four types of agents—turtles, patches, links, and an observer. Turtles move around in the NetLogo world, which is divided up in grids of patches. Each patch is a square piece of ground. Links are agents that connect two turtles. The observer does not have a location, it looks over the NetLogo world and can give commands to the patches or turtles.

Patches have coordinates. The patch at coordinates (0, 0) is called origin and the coordinates of other patches are the horizontal and vertical distances from the origin. The patch coordinates are called as pxcor and pycor like in the standard mathematical coordinates. The total number of patches is determined by the settings min-pxcor, max-pxcor, min-pycor, and max-pycor.

A turtle's position is defined by a set of coordinates called xcor and ycor. A patch's coordinates are always integers, but a turtle's coordinates can have decimals. That means a turtle can be positioned at any point within a patch; it does not have to be at its center. Links do not have coordinates, instead they have two endpoints (each a turtle). Links appear between the two endpoints along the shortest path possible even if that means wrapping around the world.

The user environment provides a panel for manipulating the turtles and a black square called the view (Fig. VI.1). The panel has three tabs—interface, information, and procedure. The interface tab is for controlling the simulation and for various displays. It allows the user to configure buttons for various actions, such as starting and running of simulations and to setup custom visualizations like graphical trends. The procedure tab opens up the screen for writing simulation programs in NetLogo language. A program usually consists of a number of routines, which are initiated either by buttons on the

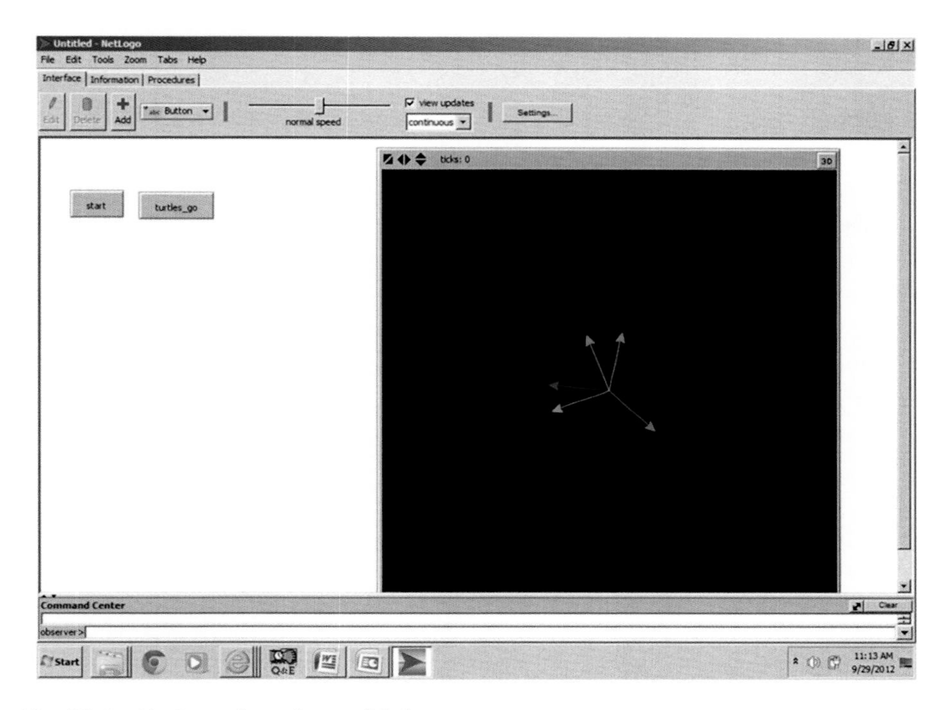

Fig. VI.1 NetLogo interface with buttons

```
to start
     clear-all
     create-turtles 5
end

to turtles_go
     ask turtles [ pen-down fd 4 ]
end
```

The first program creates 5 turtles randomly facing different directions
The second program moves them by 4 patches in the directions they are facing

Fig. VI.2 Examples of NetLogo program

information tab or by other routines. The information tab also allows a user to document the purpose and other details of a simulation in plain English.

The black square, called "View," is made up of a grid of patches (normally not visible), where the turtles move. The default shape of a turtle is a small triangle or arrowhead, which can be changed to various shapes and size to represent different agents, such as persons, automobiles, and animals. A simulation program can create these turtles and move them in any direction from patch to patch.

In the example (Figs. VI.1 and VI.2), there are two routines—start and turtles_go. The start routine clears the patch area and creates five turtles. All of them are placed at the origin, which is at the center of patch area, facing different random directions. The turtles_go routine moves each of these turtles by four patches in the directions they are facing leaving a trail of their movement.

There is an application example of agent-based modeling and simulation in Appendix VII, which is available on the NetLogo website for readers to try out.

Appendix VII: Simulation of Gender-Based Segregation Patterns

This is an example of agent-based modeling and simulation using NetLogo software package.

Anna went on a business trip to a fast developing country, which had a large potential for her company's products. She hired 40 local graduates to do marketing and sales promotions. They were young and enthusiastic, and their male/female ratio was about equal.

Anna planned to host a dinner party for these new recruits before she headed back home. Not too long ago, the country's social customs forbade social mixing of males and females, but the situation was getting more relaxed. However, she had no clue on making the seating arrangements for the dinner. A local colleague told her that having dinner on the same table with persons of opposite sex was an individual choice and the preferences varied enormously between persons.

While, some of the invitees would be more comfortable in having dinner only with members of the same gender there would be others who would actively seek the company of opposite sex. Then, there would be some in-between. As there was no easy way to know the individual preferences beforehand for making the seating arrangements, she decided to create a model to gain a better insight.

She used NetLogo package (Appendix VI) to create 40 agents, half of them female and other half male to represent the guests. As there would be eight dining tables, the agents were divided into eight equal groups at the start of a simulation run. To reflect a normal random distribution of gender preferences, each agent was randomly assigned a tolerance level of mixing with the opposite sex. The tolerance ranged from 20 through 80, where, a tolerance

A. Ghosh, *Dynamic Systems for Everyone*,
DOI 10.1007/978-3-319-43943-3

Table VII.1 Maximum of opposite gender to feel comfortable

Tolerance level	Max-opposite gender (%)
20	≤20
30	≤30
40	≤40
50	≤50
60	≤60
70	≤70
80	≤80

Male

Female

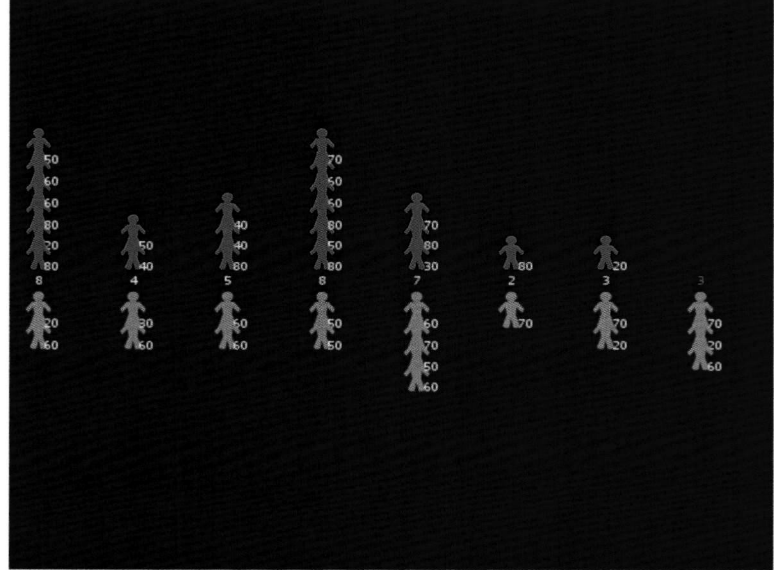

A number associated with an agent specifies its tolerance level
A number between male and female agents specifies the total agents in that group

Fig. VII.1 Guests randomly scattered at setup

level of 20 signified that the agent would be comfortable in a group in which the number of members of opposite gender would be equal to or less than 20 % (Table VII.1). Similarly, an agent with a tolerance level of 80 would be comfortable in a group in which the members of the opposite gender were equal to or less than 80 %. During simulation, the agents moved from one table to another until each of them found tables where the sex ratio was within their tolerance levels.

Anna ran the simulation a number of times and noted the number of single and mixed groups at the end of each run (Figs. VII.1 and VII.2, Table VII.2).

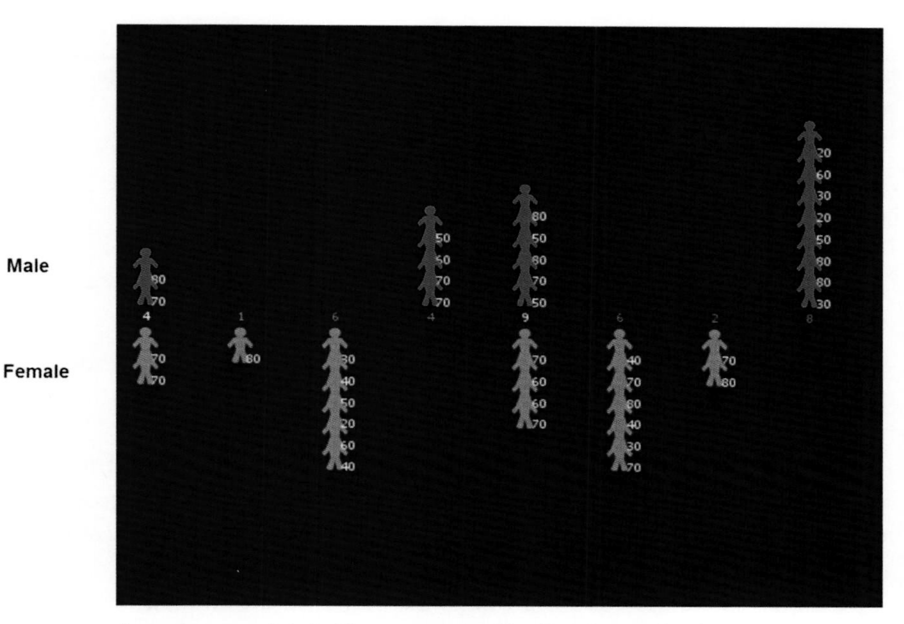

Fig. VII.2 Gender segregation after a simulation run

Table VII.2 Initial simulation results (No. of agents: 40, No. of groups: 8)

Run no.	No. of agents in single gender groups	No. of agents in mixed groups
1	40	0
2	23	17
3	23	17
4	40	0
5	40	0
Total	*166*	*34*
Percentage of total agents	*83.0 %*	*17.0 %*

The numbers varied somewhat between the runs, which was understandable because of the randomness in assigning tolerance level and the grouping of the agents at the start of each simulation. The resulting average of a number of runs dismayed Anna. Over 80 % of the agents ended up in segregated groups, leaving less than 29 % in mixed settings. It was no surprise that single gender groups were dominated by those with tolerance level of less than 50 %. What was surprising, however, was that there were a number of agents with

Table VII.3 Maximum and minimum of opposite gender to feel comfortable

Tolerance level	Max. opposite gender (%)	Min. opposite gender (%)
20	≤20	
30	≤30	
40	≤40	
50	≤50	≥0
60	≤60	≥10
70	≤70	≥20
80	≤80	≥30

greater than 50 % tolerance level in single gender groups. Anna could not quite understand why they would be in single gender groups, given that they had above 50 % tolerance for a mixed company.

That led Anna to reason that the model did not quite reflect the reality. She then modified the model by adding a function (seek_mixed_group) that made an agent with a tolerance level greater than 50 % to seek a mixed group actively.

In this mode, an agent would be comfortable in a group, where the percentage of members of the opposite gender was in a band between "Maximum Opposite Gender," and "Minimum Opposite Gender," the minimum being 50 % less than the maximum (Table VII.3). Thus, if an agent's maximum tolerance level was 70 %, then it would be happy in a group with less than or equal to 70 % and greater than or equal to 20 % of the agents of opposite gender. Anna made a number of simulation runs with this new mode and found that the number of mixed groups and single gender groups were nearly evenly divided (Fig. VII.3 and Table VII.4), which seemed to be more realistic.

Thus, she better understood the party dynamics and decided to set up eight tables, reserving half of them for those who prefer mixed company. The other four tables were divided equally for male and female guests. She kept a set of extra chairs handy to take care of possible imbalances.

The Two Main Lessons Anna Learned from This Exercise

- Small levels of preference or restrictiveness can unwittingly lead to large levels of segregation.
- A small amount of intentional action, like seeking a mixed group, even without altering the preference levels will allow opportunities for interaction within a wider range, leading to a more balanced situation.

Readers may try out the simulation of "Party Segregation," which is in the NetLogo community model website: http://ccl.northwestern.edu/netlogo/models/community/Party%20Segregation.

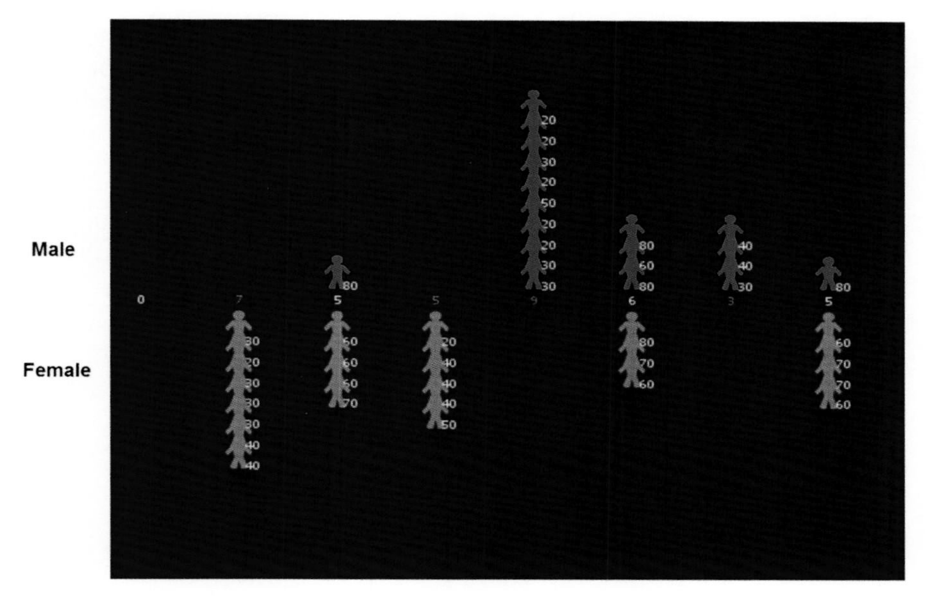

Those with tolerance level greater than 50% seek mixed groups

A number associated with an agent specifies its tolerance level
A number between male and female agents specifies the total agents in that group

Fig. VII.3 Lesser gender segregation when seeking mixed group

Table VII.4 Simulation results with "seek_mixed_group" function on (No. of agents: 40, No. of groups: 8)

Run no.	No. of agents in single gender groups	No. of agents in mixed groups
1	22	18
2	16	24
3	12	28
4	20	20
5	22	18
Total	*92*	*108*
Percentage of total agents	*46.0 %*	*54.0 %*

In the USA, city neighborhoods are often highly segregated along racial and ethnic lines. It may appear that this is due to high levels of intolerances and self-segregation. However, Thomas Schelling, a famous economist, showed that a small preference for one's neighbors to be of the same color often leads to total segregation. In many cases, a little remedial action by the town authorities, such as a modest property tax breaks and providing social amenities like playgrounds and community centers to mixed neighborhoods, may go a long way in addressing such imbalance.

Appendix VIII: Documentation of Procedural Functions

There are a number of ways of documenting procedural functions. Some of them like sequential function chart and procedure function chart are defined by international standards. While others, such as flowchart and Structured English are not, but are widely used. Choosing a documentation method depends on the application and user preference. For common understanding, it is preferable to use a single documentation method within an organization.

The Fundamental Constructs in a Procedure

A set of discrete events in a procedure, whether simple or complex may be expressed by a combination of a set of simple constructs (Fig. VIII.1), which are

- Action command
- Conditional branching
- Parallel operation
- Repetitive looping

Action command, as the name implies is an instruction to perform a task. That may be to adjust the set point of a heater to right temperature or to break a shell containing nuts with a stone. An action command may also be used to invoke another procedure, such as "Charge Water" or "Store."

Conditional branching allows invoking a set of logic among multiple sets depending on certain current conditions. It is like choosing a path at a fork

A. Ghosh, *Dynamic Systems for Everyone*,
DOI 10.1007/978-3-319-43943-3

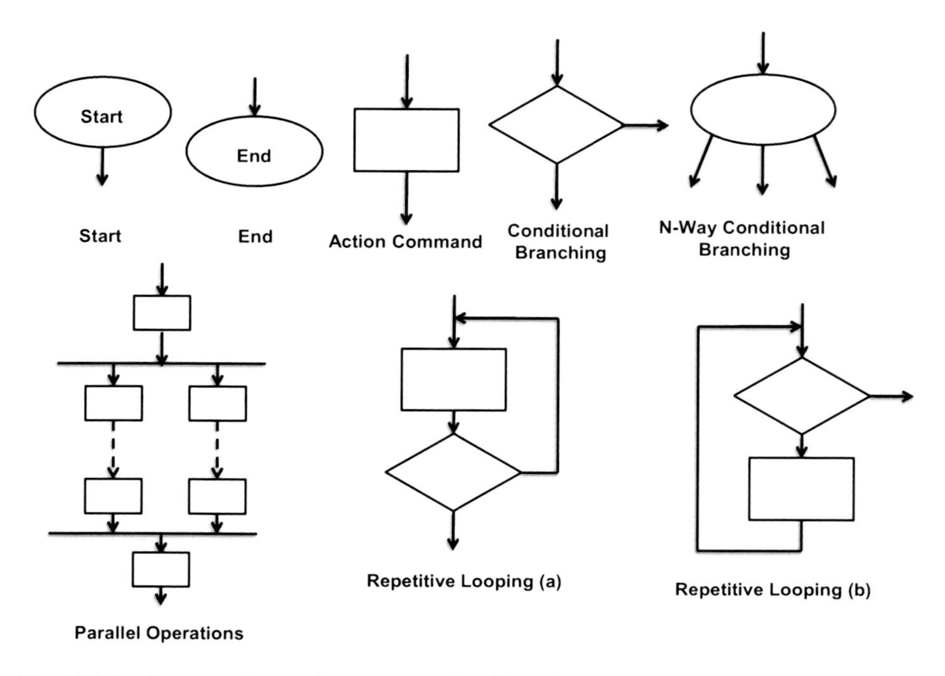

Fig. VIII.1 Commonly used constructs for flowcharts

on the road. For example, store a product if it passes quality test; otherwise discard it to the recycle bin.

Parallel operations are those that may run simultaneously, such as charging a raw material to a reactor along with the charging of water.

Repetitive looping allows repeating one or more steps in a procedure until certain conditions are met. A capuchin monkey, for example, may try to repeat the procedure for breaking a shell for a given number of times and leave that to rot if not successful. Repetitive looping function may be viewed as a combination of action commands and conditional branching.

A procedural function may be documented in many different ways—graphical, tabular, textual, or their combinations. Following paragraphs elaborate some of the commonly used documentation procedures.

Flowchart

Flowchart is a popular method for documenting procedures. It is graphical and is fairly easy to learn and understand. Each action-command step in a flowchart is represented by a rectangular box (Fig. VIII.1). Arrows show the

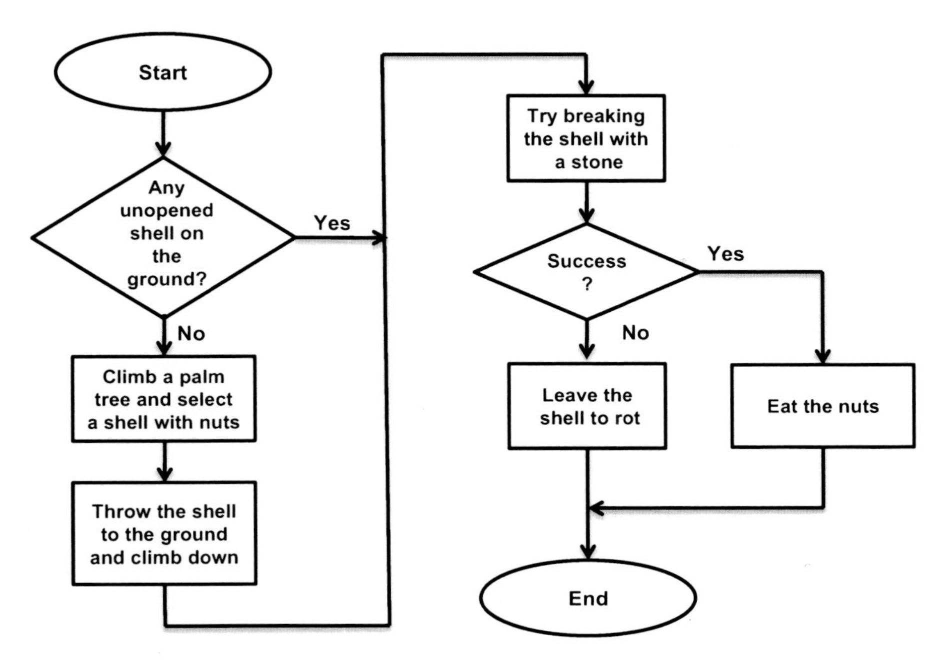

Fig. VIII.2 Flowchart for fetching, cracking, and eating nuts

connections between steps and the direction of procedural flow. The start and end of a procedure are usually represented by ovals. A diamond represents a two-way conditional branching and an oval may represent a multiple conditional branching, with an arrow for each possible branch.

Single horizontal lines are used to indicate the start and end of a parallel operation. A combination of rectangular boxes and diamonds are used when, one or more steps need to be repeated. The condition for repetition may be specified either at the beginning or at the end of the step(s) that need to be repeated.

The procedure for fetching, cracking, and eating nuts is from an example of a monkey's quest for nuts as described in Chap. 8 (Fig. VIII.2).

A set of flowcharts may be drawn in a hierarchical fashion, where a step in a flowchart at a higher level may represent the entire function of a flowchart at a lower level. A flowchart is easy to draw and understand when it is limited to one sheet of paper, as jumps and loops are obscured when spread around multiple sheets. Modularization of a long and complicated procedure is of great help in reducing complexity.

Though flowchart is in use for a long time, very little effort has been made towards its standardization. Thus, there are many different flowchart dialects in use.

Sequential Function Chart

Sequential function chart (SFC) was designed to specify discrete steps in a machine control or discrete part manufacturing (Fig. VIII.3). In SFC, the transition from one step to the next is explicitly stated, with a single horizontal line and the start and end steps are depicted by rectangles with double lines. Branching is specified by a single horizontal line whereas double horizontal lines specify parallel operations.

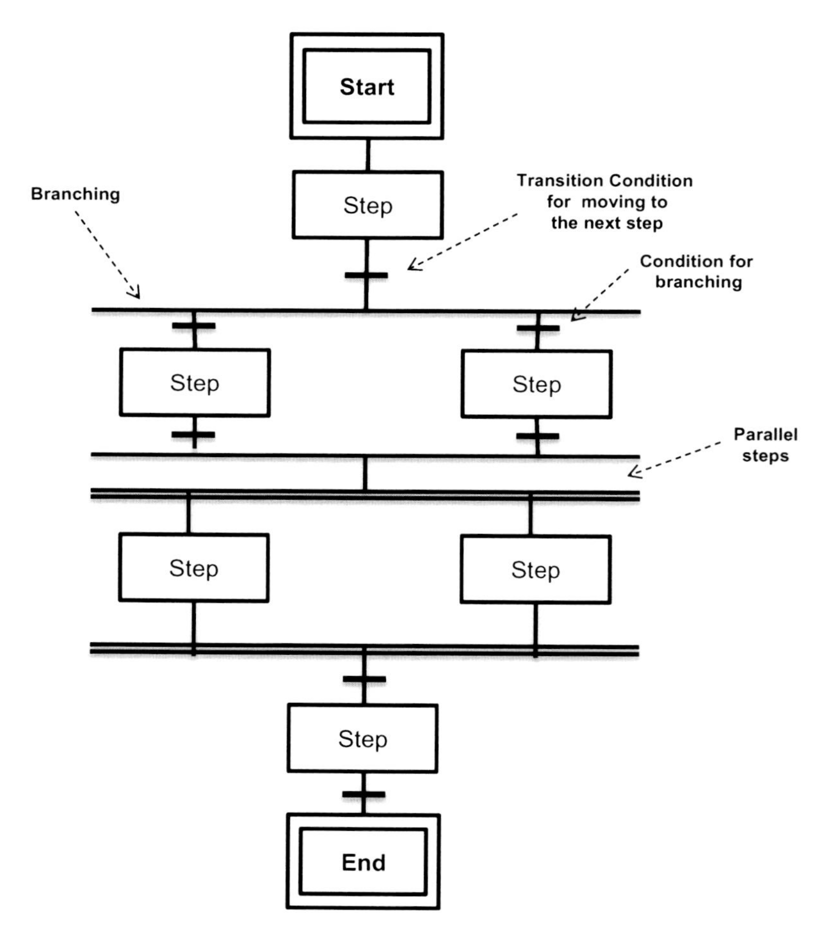

Fig. VIII.3 Sequential function chart

Procedure Function Chart

Procedure function chart (PFC) was derived from SFC, to depict procedural functions in process industries, such as in the manufacture of pharmaceuticals or fine chemicals.

The PFC is defined by IEC (International Electrotechnical Commission) standard 61131–3. Some of the PFC rules are described here (Figs. VIII.4 and VIII.5).

Action Command An action command is a step in a procedure, which is represented graphically by a rectangular block containing a step name or a number. Vertical lines at the top and bottom of a step specify the links to and from the step. One or more actions may be associated with each step. An action may be specified by a function name or a set of instructions. A "+" sign at the upper right corner of an action command indicates that the step invokes a procedure at a lower level.

The start and end steps are represented by triangles. In an automated system, the start step is activated manually or by another procedure and a following step is activated on completion of the preceding step(s).

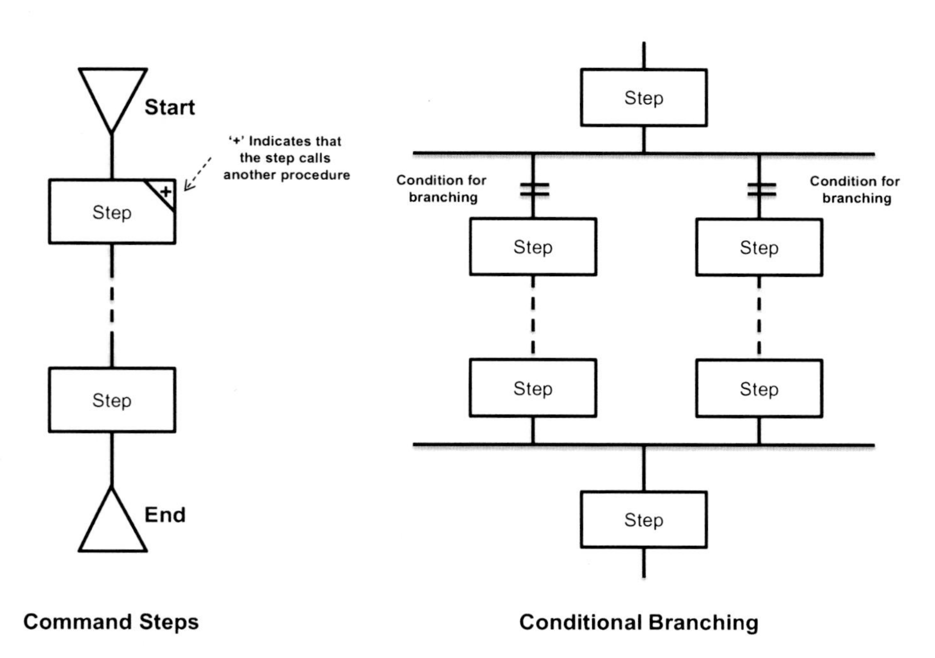

Fig. VIII.4 Procedure function chart constructs (1)

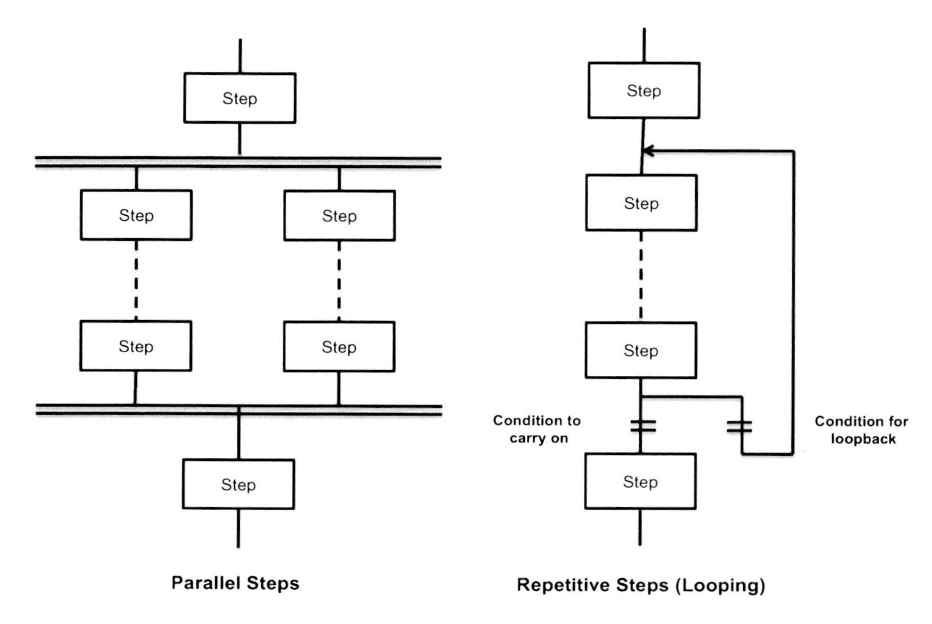

Fig. VIII.5 Procedure function chart constructs (2)

Conditional Branching Conditional branching is specified by a single horizontal line along with the conditions for branching to a set of steps. There is no set limit on the number of conditional branches that may be specified. A specified condition is usually expressed as a logical function, which needs to be satisfied to activate the step that follows. A single horizontal line may also indicate an end of conditional branching.

Parallel Steps Parallel steps are specified by using double horizontal lines, where any number of steps may run in parallel. Double horizontal lines may also indicate the end of a parallel operation.

Repetitive Looping Normally, the steps run sequentially from the top to the bottom, where no arrows are shown. However, where one or more steps have to be repeated for a number of times a line with an arrow is usually used to specify the loopback path. There, the conditions for looping back or for continuing forward are specified.

Transition Condition The transition condition from one step to the next is generally implicit. That is, the completion of a step leads to the start of the next step(s). Explicit transition, where needed, is specified by a pair of little horizontal bar with a statement or an equation specifying the condition(s), such as:

- Passed quality test
- Outside temperature > 75 °F

PFCs offer a number of advantages in depicting sequential functions. It is easy to learn and understand and parallel operations can be shown clearly. As it follows international standard, the ambiguity in drawing and interpreting a PFC is significantly reduced.

Structured English

Structured English is another method for documenting procedures. In Structured English, the procedure steps are represented by English narrative written within a set of rules of structure and style to keep them clear, unambiguous, and at the same time quite detailed. Structured English is akin to standard English language, which makes it easy for a layperson to follow. Following are some of the rules for writing a procedure in Structured English.

A Structured English narrative is composed of statements that may be divided into two broad types—non-executable and executable. A non-executable statement does not specify any action, whereas executable statements specify action commands, conditional branching, and repetitive looping. Headers, labels, and comments are examples of non-executable statements. The example for fetching and cracking nuts by capuchin monkeys in Brazil may be specified in Structured English (Fig. VIII.6), where the procedure title (header) and the labels Start, End, and Break-the-Shell are non-executable statements. In this example, If, Then, and Else statements are used to specify conditional branching and those without them are action commands.

In Structured English, a statement, either non-executable or executable, should start on a new line. The statements should be properly indented to show the blocks of logic that go together. The details of rules and syntax vary between applications, as there is no standard or normative way to specify in Structured English. Therefore, when a group is involved in specifying and interpreting a set of procedures, they need to agree upon the rules and conventions they would use for Structured English. Following points will help in developing an easy to understand Structured English procedure:

- Document each statement on a new line.
- Break up a procedure into modules, so that each module requires no more than one page to document.
- Use simple statements and avoid negative logic.

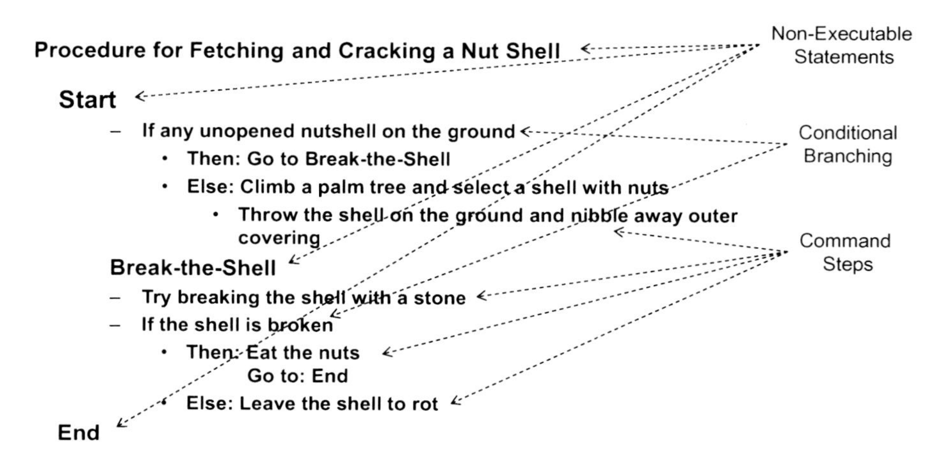

Fig. VIII.6 Procedure for fetching, cracking, and eating nuts in Structured English

- Indent properly to identify a group of related functions.
- Avoid excessive nested logic.

A procedure in Structured English is easy to generate, as it requires only a word processor in a personal computer. If properly modularized, it is easy to understand and modify. In many engineering applications, Structured English can be directly compiled into computer code for automatic execution, where direct compilation is not possible, computer code may be directly embedded into a procedure, where there may be one-to-one correspondence between a statement and the code that follows. Thus, making the code easy to understand, debug, and modify.

The main drawback of Structured English is its lack of generally accepted rules and guidelines. A procedure in Structured English can be obscure for a long and complicated piece of logic, and parallel operations are not easy to document. Proper modularization can, however, reduce or eliminate these problems.

Appendix IX: Example of a Simple Batch Process

This is an example of the use of procedural functions for manufacturing chemicals in batches. It shows how a recipe with a set of phases specifies the process. It illustrates how the same procedure can be used to manufacture different grades of products by changing a set of variables. It shows the different states in a manufacturing process and shows how a monitoring function, on detecting a malfunction, initiates an exception procedure to take care of the problem.

By studying the underlying principles of such a process, we are able to apply similar methods for analyzing and designing procedural functions for social, legal, political, and other such systems.

Process Description

Green Rock Chemical Company is the manufacturer of biodegradable SuperX solvent, which is widely used for making detergents and household cleaners. The solvent is manufactured in batches by the reaction of two raw materials, RM1 and RM2 in an aqueous solution. The temperature of the solution is controlled within close limits during the reaction to maintain the quality of the product (Fig. IX.1).

There are two tanks for holding raw materials and a weigh tank for weighing them accurately before charging them into the reactor. A flow totalizer measures the amount of water charged to the reactor. The reactor is heated by circulating hot water in a jacket around the reactor.

A. Ghosh, *Dynamic Systems for Everyone*,
DOI 10.1007/978-3-319-43943-3

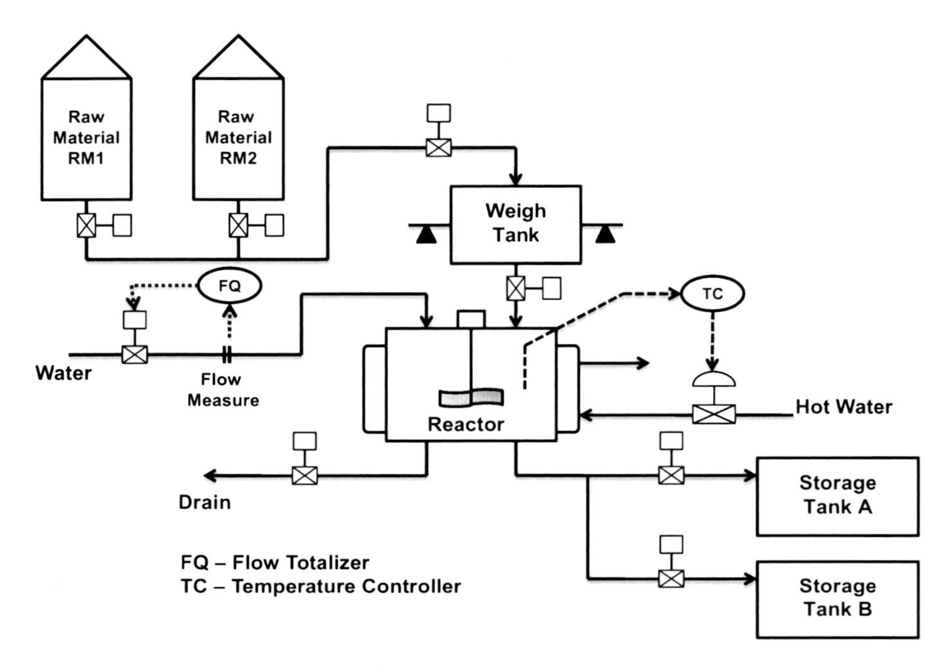

Fig. IX.1 SuperX solvent manufacturing plant

After charging right amount of water and the raw materials, the contents in the reactor are mixed by running the agitator. The reactor contents are heated up to a specified temperature to start the reaction. On completion of the reaction, the product is stored in one of the storage tanks, if its quality is found to be within accepted limits. If the product does not meet the required quality specification, it is drained to a sump for recycling.

The chemical company manufactures two grades of solvents, SuperXL (light) and SuperXP (premium), where the manufacturing procedures for these two grades are quite similar.

Recipes and Phases

The top-level procedure for manufacturing a product, such as SuperX solvent, is called a "Recipe" (Fig. IX.2). The recipe invokes a number Phases that perform various tasks, such as, Charge Water, Charge Raw Material, React, and Drain. Each of these phases is a procedure for performing a specified task, for example, Charge Water for supplying a specified amount of water to the reactor.

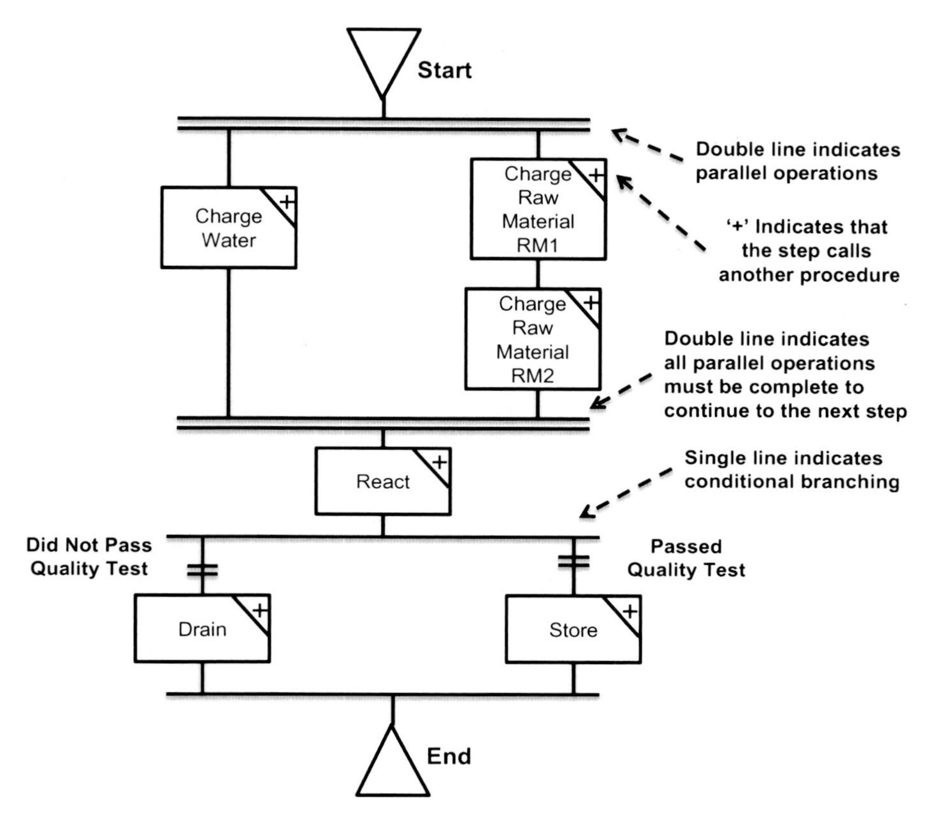

Fig. IX.2 Recipe procedure (*top level*) for manufacturing SuperX solvent

A phase consists of a number of instruction steps, where each instruction specifies a major process step. In the phase Charge Water, for example, some of the major steps are—Initialize Flow Totalizer and Open Water Inlet (Fig. IX.3). Notice that each step calls for a specific set of actions for manipulating relevant parts of the manufacturing plant. Similarly, the React phase (Fig. IX.4) invokes a number of process steps to carry out the reaction for producing the SuperX solvent.

The example specifies only two levels of procedures. However, there are no limits on the number of levels or the number of procedures in each level. Except for one at the highest level, where there is only one procedure, which is called a Recipe (Fig. IX.5).

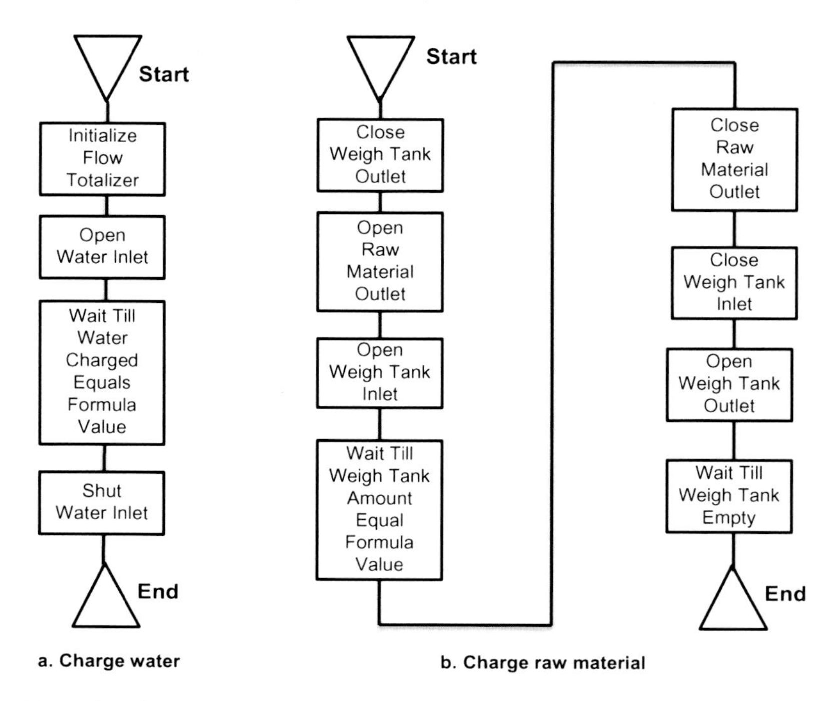

Fig. IX.3 The charging phases

Common Procedure for Charging Raw Materials

You may notice that there is only one procedure for charging the two raw materials RM1 and RM2 (Fig. IX.3b). That is because the procedures for charging the two raw materials are essentially similar, differing only in the amount of raw materials to be charged. When the recipe calls for charging of one of the raw materials, it specifies whether it is RM1 or RM2. The phase for charging a raw material picks up the material type and acts accordingly.

Common Recipe with Different Formula Variables In this example, only one recipe has been specified even though two different grades of solvents SuperXL and SuperXP are manufactured. That is possible by specifying two sets of variables, one for each grade of the product. A set of such variables for manufacturing a product is called a Formula table (Table IX.1). Thus, when the SuperXL is manufactured, 500 gallons of water is charged in the reactor along with 50 gallons of raw material 1 and 75 gallons of raw material 2. While for SuperXP, they are 450 gallons, 75 gallons, and 100 gallons, respec-

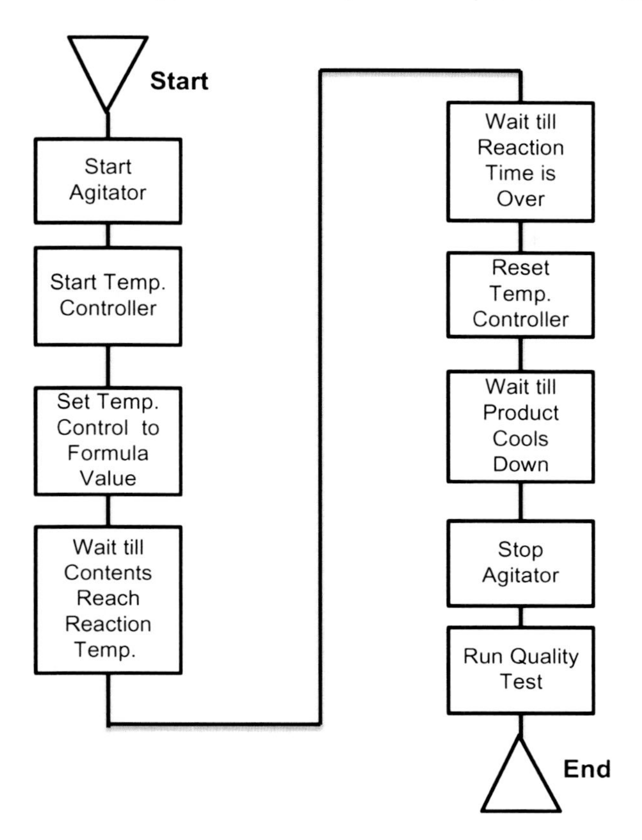

Fig. IX.4 Phase: React

tively. Similarly, the reaction time and temperature are changed according to the product. Finally, there are separate storage tanks for each of these products, which are also specified in the formula table.

Emergency Shutdown Procedure

There are many possible reasons for aborting a manufacturing process. Those include system problems, such as equipment failure, line blockage, and control loop malfunction. Additionally, failure of water supply or fire may require shutting down the process. In an automated system, one or more monitoring functions watch the plant and the process to ensure that it is running normally. If a failure condition occurs, then the monitoring function invokes exception procedure(s) (Fig. IX.6).

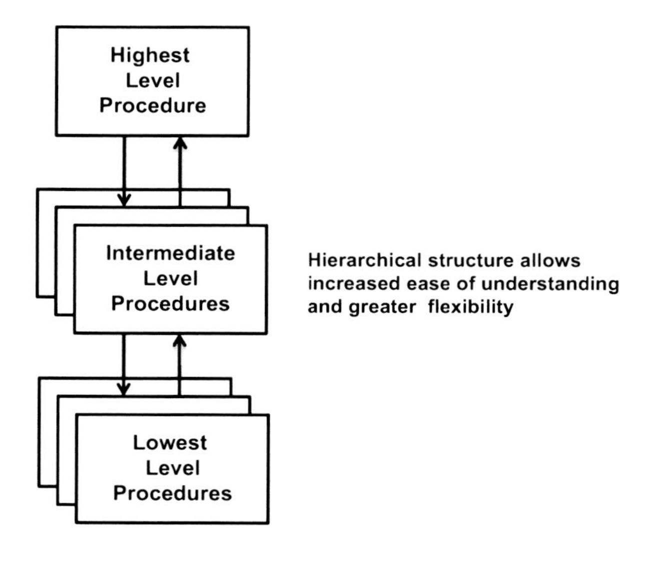

Fig. IX.5 Hierarchy of procedures

Table IX.1 The formula variables

	Variable	SuperXL	SuperXP
1	Water	500 Galls	450 Galls
2	Raw material 1 (RM1)	50 Galls	75 Galls
3	Raw material 2 (RM2)	75 Galls	100 Galls
4	Reaction temp	175 °F	190 °F
5	Reaction time	3.00 h	4.00 h
6	Reactor cool down temp	75 °F	75 °F
7	Product storage tank	A	B

The exception procedure may be a shutdown procedure, where the supply of raw materials to the reactor is terminated, heating the reactor is stopped, and the contents of the reactor are drained after it has cooled down. There may also be parts of a shutdown procedure that take specific actions to maintain safety—like initiating fast cooling of the reactor, rather than simply shutting off the heating system. In a more complex manufacturing process, there could be multiple shutdown procedures for the different stages of a manufacturing process or for different reasons of failure.

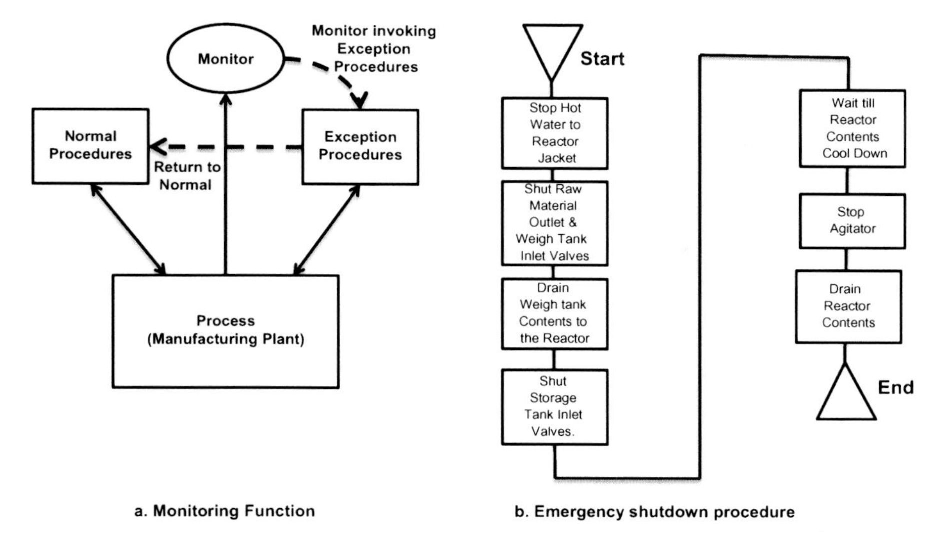

Fig. IX.6 Monitoring function and emergency shutdown procedure

The States of a Procedure

Unlike a continuous function, which is generally active most of the time, a procedural function has a start and an end. Between the start and the end, there could be a number of states, such as idle, running, paused, held, stopped/aborted, and complete (Fig. IX.7).

These states may be defined as:
Idle: The starting procedure is ready, but waiting for a command to run.
Running: A procedure is running normally.
Paused: The normal running of a procedure has been temporarily halted because of a problem. It will run again, when the problem is resolved, starting at the point where it stopped.
Held: The normal running of the procedure has been halted because the problem is more serious, which causes an exception logic to run. The exception logic is either able to take care of the situation or decides to abort the batch.
Restarted: This state will transition to normal running after the exception logic has taken care of the problem.
Aborted: This state is reached when the problem cannot be rectified by the exception logic. At this state, the batch of chemical is discarded and the procedure reverts to idle state.

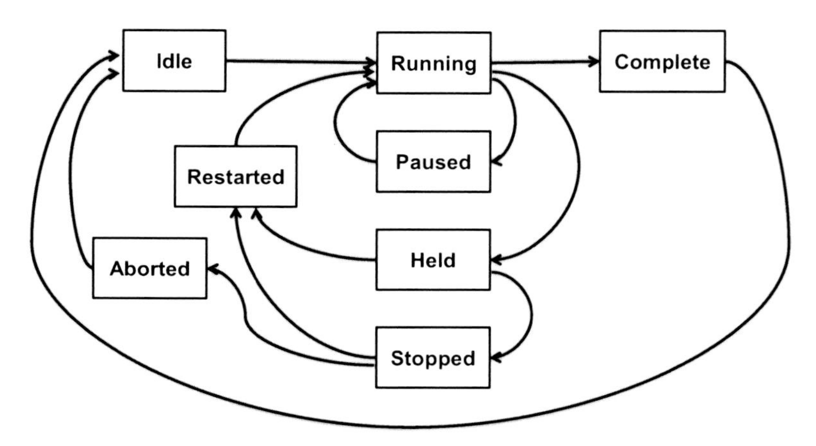

Fig. IX.7 Procedure states (a simplified view)

Complete: This state is reached when a whole set of the procedures have run, producing the necessary product. The next logical state is idle.

Additional Comments

It should be noted that a set of procedures might be used for manufacturing a product irrespective of whether it is executed manually, automatically.

In this example, there is a top-level recipe procedure, which activates other procedures called phase logic to perform a set of functions. A phase logic may directly manipulate plant equipment, such as valves or control loops. It may also set continuous control parameters. For example, it may activate a temperature controller after setting its set point.

Proper documentation of a procedure has many benefits. The very exercise of putting thoughts on paper leads to better analysis and understanding of a process. A well-documented procedure leads to a uniform level of understanding among a group, thus reducing misunderstanding and misinterpretation.

A procedure is like a well-proven model in a model-based control. Procedures do allow decision points that make it possible to choose from alternate sets of actions depending on current process conditions. As in the case of storing or rejecting a product near the end of reaction or adjusting one or more variables based on quality tests during a manufacturing process. Normally a procedure does not require any major modification or update during its execution. That, however, does not preclude the updating of a procedure at the end of its execution to alter or improve its performance in subsequent runs.

Terminology

Action generator A part of a controller that generates the signal for controlling a variable

Agent An intelligent object that acts on its own based on its environmental conditions

Analog Representation of a piece of information by its physical quantity

Artificial neural network A model that emulates the biological neural system in a brain

Balanced state A condition where the internal and external signals in a system balance each other (synonymous to Steady State)

Balancing loop A loop where variables counteract (negative feedback) thus creating a balanced state

Batch process A process for manufacturing a product in batches

Biofeedback A process of gaining greater awareness of bodily functions primarily by using sensors

Black box model A model that is generated by observing the changes in the inputs in a system and their resulting outputs

Carbon dioxide A gas composed of two oxygen atoms a single carbon atom. It is a greenhouse gas

Carbon monoxide A gas composed of one oxygen atom and one carbon atom. It is toxic to human and other animals

Cascaded control A configuration where the output of one controller is the set point of another

Conceptual model A model that shows the main functions of a system, without getting into great details

Constraint analysis Finding the barriers within which a system should operate

A. Ghosh, *Dynamic Systems for Everyone*,
DOI 10.1007/978-3-319-43943-3

Dead time The time interval between an action and the start of its observable result (see Lag)

Derivative control Where the output of a controller is proportional to the rate of change the error

Detail complexity Complexity due to a large number components or subsystems but they do not change with time

Digital Representation of a piece of information by its numerical value

Discrete event An event that does not happen continuously but occurs at a certain time or due to certain conditions

Distributed intelligence The behavior of a system where a large part of its decision-making process is distributed in its various subsystems

Dynamic complexity Complexity due to changes in the behavior of a system with time or other conditions

Dynamic state A condition where one or more system variables change significantly over a period of time

Encapsulation A technique for hiding a complex function inside an object

Engineered system Human designed mechanical, electrical, and other such systems

Error The difference between a process output and its set point; usually varies with time

Exponential growth The growth rate of a variable that is proportional to its current value

Feedback Information about the state of a system, usually after an action that caused some changes

Formula A set of data that a recipe needs to manufacture a batch of product

Gain The ratio between the change in output to a change in input; thus, higher the gain higher is the output for a given input.

Gas chromatograph A device that measures the composition of a mixture of chemicals

Goal seeking The behavior of a system when it tries to reach a goal or a set point

Greenhouse gas Gases that trap heat and make the planet warmer. The primary greenhouse gases are carbon dioxide, water vapor, methane, nitrous oxide, and ozone.

Hill climbing A technique for optimizing a system by taking multiple steps

Holism A view that a system functions as a whole and its behavior cannot be fully understood by solely looking at the properties of its component parts

Inheritance A function that allows the characteristics of an object to be copied or inherited by another object

Integral control Where the output of a controller is proportional to a sum of the errors

Interactive model A model whose inputs can be manipulated to observe the resulting changes in its outputs

Kaizen A method of continuous improvement in small steps

Lag The time interval between an action and the resultant change reaching nearly 2/3 of its final value; often called as the time constant of the system (see Dead Time)

Late binding The process of establishing links between objects as and when needed rather than setting them beforehand

Linear growth The growth rate of a variable that follows a straight line

Mental model A model that resides in a brain

Message passing Requesting an object to carry out its function by sending a message

Model The descriptions or representations of something that is material or abstract

Multivariable control Controlling a system with multiple interactive variables

Object A software module that include program and data, which often mimics a real life entity

Object oriented programming A computer programming method that uses objects rather than subroutines

Offset Similar to error but usually more persistent

Optimization The process of finding the best possible solution for a given problem

Oscillation Repetitive variation of a variable around a central value

Ozone A highly reactive form of oxygen

Phase A major part of a recipe

Procedure An ordered set of instructions that specify a function or a task to be performed

Proportional control Where the output of a controller is proportional to the error

Recipe A procedure for manufacturing a product in batches

Reductionism The view that a system's functions can be better understood by closely studying the behavior of its component parts

Reinforcing loop A loop, where the variables reinforce each other leading to increased growth or decline

Sensor An element or a device that measures the value of a variable

Servomotor A motor for precise control of angular position, velocity, and acceleration

Set point The desired value of a variable that is being controlled

Steady state A condition where the system variables show little change over a period of time

Swarm intelligence The collective behavior of a larger number of self-organized systems

System A set of connected things or parts; an organized body of material or immaterial things

System Science A field of study on the nature of systems

Thermostat A device that controls the temperature of a system, such as a room or a building

Time invariant Where a system produces the same response to an input regardless of time

Time variant Where a system's response to an input varies with time

Variable A thing or a value that is changeable

Watt governor A device for controlling the speed of an engine.

Acronyms

AI	Artificial intelligence
ANN	Artificial neural network
CD	Certificate of deposit
CO_2	Carbon dioxide
DARPA	Defense Advanced Research Projects Agency (USA)
DNA	Deoxyribonucleic acid
DPM	Dynamic performance measures
EEG	Electroencephalogram
EPA	Environmental Protection Agency
fMRI	Functional magnetic resonance imaging
GPS	Global positioning system
HR	Human resources
IEC	International Electrotechnical Commission
ISA	International Society of Automation
MAS	Multi-agent system
MBC	Model-based control
MIT	Massachusetts Institute of Technology
MRI	Magnetic resonance imaging
MTBE	Methyl tertiary butyl ether
NO_2	Nitrogen dioxide
NSAID	Nonsteroidal anti-inflammatory drug
OOP	Object-oriented programming
PTSD	Post-traumatic stress disorder
REM	Rapid eye movement; a type of sleep pattern
rtfMRI	Real-time functional magnetic resonance imaging
SARS	Severe acute respiratory syndrome
SID	Sudden infant death
UNICEF	United Nations International Children's Emergency Fund

A. Ghosh, *Dynamic Systems for Everyone*,
DOI 10.1007/978-3-319-43943-3

Bibliography

Chapter 1

1. De Latil, P. (1957). *Thinking by machine*. Boston: Houghton Mifflin Company.
2. Ford, A. (2010). *Modeling the environment*. Washington: Island Press.
3. Frisman, P. (2003). *Connecticut government research report*. http://www.cga.ct.gov/2003/olrdata/env/rpt/2003-R-0911.htm. Accessed 24 Nov 2014.
4. Hoppensteadt, F. Predator–prey model. http://www.scholarpedia.org/article/Predator-prey_model. Accessed 24 Nov 2014.
5. Popham, P. (1998). *Arsenic-tainted water from UNICEF wells is poisoning half of Bangladesh*. London: The Independent. http://www.independent.co.uk/news/arsenictainted-water-from-unicef-wells-is-poisoning-half-of-bangladesh-1196091.html. Accessed 24 Nov 2014.
6. Von Bertalanffy, L. (1969). *General system theory: foundations, development, applications*. New York: Penguin University Books.

Chapter 2

7. Dynamic Performance Measurement System Improves Energy and Electricity Consumption. (2009). *Invensys brochure*. Foxborough: Invensys.
8. Encyclopedia Britannica. (1975). James Watt, Macropaedia (Vol. 19, page 662).
9. Fiske, T. (2000). Real-time accounting, automation world, Chicago, IL. http://www.automationworld.com/plm/real-time-accounting. Accessed 24 Nov 2014.
10. Martin, P. G. (2006). *Bottom line automation*. Research Triangle Park: ISA.
11. Meadows, D. H. (2008). *Thinking in systems*. White River Junction: Chelsea Green Publishing Company.

A. Ghosh, *Dynamic Systems for Everyone*,
DOI 10.1007/978-3-319-43943-3

Chapter 3

12. Dembosky, A. (2011). *Invasion of the body hackers*. New York: Financial Times. http://www.ft.com/intl/cms/s/2/3ccb11a0-923b-11e0-9e00-00144feab49a.html#axzz1Tl5847rc. Accessed 24 Nov 2014.
13. Fisher, L. M. (2005). The prophet of unintended consequences. Booz & Co., Inc. http://www.strategy-business.com/media/file/sb40_05308.pdf. Accessed 24 Nov 2014.
14. Forrester, J. W. (1969). *Urban dynamics*. Forrester: Pegasus Communications, Inc.
15. Haraldsson, H. V. (2000). *Introduction to systems and causal loop diagrams*. Lund: Lund University. http://dev.crs.org.pl:4444/rid=1244140954250_1167059429_1461/Introduction%20to%20Systems%20and%20Causal%20Loop%20Diagrams.pdf. Accessed 24 Nov 2014.
16. Origin of System Dynamics. http://www.systemdynamics.org/DL-IntroSysDyn/origin.htm. Accessed 24 Nov 2014.
17. Peterson, S., Bridgewater, K., et al. (2011). *Youth violence systems project (Special Edition Review)*. Boston: Emmanuel Gospel Center.
18. Richardson, G. P. (1986). *Problems with causal-loop diagrams revisited, system dynamics review*. Albany: System Dynamic Society. http://www.clexchange.org/ftp/documents/system-dynamics/SD1997-09ProblemsInCLDsRevi.pdf. Accessed 24 Nov 2014.
19. ScienceDaily. (2009). New findings on Parkinson's disease and effect on patient behavior. http://www.sciencedaily.com/releases/2009/06/090630163148.htm. Accessed 24 Nov 2014.

Chapter 4

20. Chernobyl Accident 1986. (2014). World Nuclear Association. http://www.world-nuclear.org/info/Safety-and-Security/Safety-of-Plants/Chernobyl-Accident/. Accessed 24 Nov 2014.
21. Dorner, D. (1996). *The logic of failure*. New York: Basic Books.
22. Global Research. (2008). *Nikolai Kondratiev's "long wave": the mirror of the global economic crisis*. http://www.globalresearch.ca/nikolai-kondratiev-s-long-wave-the-mirror-of-the-global-economic-crisis/11161. Accessed 24 Nov 2014.
23. Senge, P. M. (2006). *The fifth discipline*. New York: Doubleday/Random House.

Chapter 5

24. Ford, A. (2010). *Modeling the environment*. Washington: Island Press.

25. Gershenson, C. Artificial neural networks for beginners. http://arxiv.org/ftp/cs/papers/0308/0308031.pdf. Accessed 24 Nov 2014.
26. Halford, G. S., Baker, R., McCredden, J. E., & Bain, J. D. (2005). How many variables can humans process? http://www.ncbi.nlm.nih.gov/pubmed/15660854. Accessed 24 Nov 2014.
27. Peterson, S., Bridgewater, K., et al. (2011). *Youth violence systems project (Special Edition Review)*. Boston: Emmanuel Gospel Center.
28. Starting Point Teaching Entry Level Geoscience. Why are models useful? Carleton College, Northfield, MN. http://serc.carleton.edu/introgeo/models/Usefulness.html. Accessed 24 Nov 2014.
29. Wong, G., & Wong, V. The training of astronauts: virtual reality in space exploration. http://www.doc.ic.ac.uk/~nd/surprise_96/journal/vol4/kcgw/report.html. Accessed 24 Nov 2014.

Chapter 6

30. Brooks, F. P. (1995). *The mythical man month*. Boston: Addison Wesley.
31. Considine, D. M. (1985). *Process instrument and controls hand book*. New York: McGraw-Hill.
32. Kaizen. Wikipedia. http://en.wikipedia.org/wiki/Kaizen. Accessed 24 Nov 2014.
33. Kepner & Tragoe. http://www.kepner-tregoe.com/. Accessed 24 Nov 2014.
34. Rhinehart, R. R. (2006). *Model based control. Instrument engineers handbook (Vol. II)*. Boca Raton: Taylor & Francis.
35. Shimizu, K. Transforming Kaizen at Toyota. Okayama University, Japan. http://www.e.okayama-u.ac.jp/~kshimizu/downloads/iir.pdf. Accessed 24 Nov 2014.

Chapter 7

36. Agent Builder. (2004). Auction agents for the electric power industry. Research conducted for the Electric Power Research Institute, Palo Alto, CA. http://www.agentbuilder.com/Documentation/EPRI/. Accessed 24 Nov 2014.
37. Chiu, S., Yi-Liang, C., et.al. (2003). Distributed diagnostics & reconfiguration for shipboard chilled water system; Thirteenth International Ship Control Systems Symposium, Orlando, Florida. http://chius.homestead.com/files/DiagReconfig.pdf. Accessed 24 Nov 2014.
38. Ghosh, A. (1991). *Programming with modules*. Chemical Engineering, New York: McGraw Hill.
39. Ghosh, A., & Woll, D. (2006). *Intelligent agents deliver robust and flexible automation in dynamic environments, ARC Strategies*. Dedham: ARC Advisory Group.
40. Gilbert, N. (2008). *Agent-based models*. Los Angeles: Sage.

41. Jennings, N. R., & Bussmann, S. (2003). *Agent-based control systems.* IEEE Control System Magazine.
42. Miller, P. (2010). *The smart swarm.* New York: Avery.
43. Pipattanasomporn, M., Feroze, H., & Rahman, S. (2009). Multi-agent systems in a distributed smart grid: design and implementation. Proceedings of IEEE PES 2009 Power Systems Conference and Exposition, Seattle, WA.
44. Rosenof, H. P. (1998). *A spot market for electricity.* Cambridge: Gensym Corporation.

Chapter 8

45. Borshchev, A., & Fillippov, A. (2004). *From system dynamics and discrete event to practical agent based modeling: Reasons, techniques, tools.* Oxford: International Conference of the System Dynamics Society.
46. Burns, K., & Novick, L. Prohibition: a film by WGBH Boston. http://www.pbs.org/kenburns/prohibition/unintended-consequences/. Accessed 9 May 2016.
47. Craig, L. (2011). Nature of procedure. Paper presented at WBF Conference, Newark, Delaware.
48. Ghosh, A., & Rosenof, H. (1987). *Batch process automation: theory and practice.* New York: Van Nostrand Reinhold.
49. Ghosh, A. (2012). *Batch process automation, instrument engineer's handbook* (Vol. III). Research Triangle Park: ISA; Boca Raton: CRC Press.
50. IEC 61131–3. (2013). Programming languages standard for programmable controllers. International Electrotechnical Commission, Geneva, Switzerland.
51. IEC 61512–1 (1997) & ANSI/ISA-88.00.01. (2010). Batch control standard part 1: models and terminology. International Electrotechnical Commission, Geneva, Switzerland and The International Society of Automation, Research Triangle Park, NC.
52. IEC 61512–2, & ISA-88.00.02. (2001). Batch control standard part 2: data structures and guidelines for languages. International Electrotechnical Commission, Geneva, Switzerland and The International Society of Automation, Research Triangle Park, NC.
53. Wikipedia, Cobra effect. https://en.wikipedia.org/wiki/Cobra_effect. Accessed 9 May 2016.

Chapter 9

54. Biswas, A. K. Aswan Dam revisited: the benefits of a Much-Maligned Dam. http://www.icid.org/aswan_paper.pdf. Accessed 24 Nov 2014.

55. Forrester, J. W. (1971). Counterintuitive behavior of social systems. *MIT Technology Review, 73*(3), 52–68.
56. Harper, J. (2010). *Excessive cleanliness may boost allergies*. The Washington Times, http://www.washingtontimes.com/news/2010/apr/15/excessive-cleanliness-may-boost-allergies/. Accessed 24 Nov 2014.
57. Hart, G. (2003). *Science sometimes has unintended consequences*. http://www.geoff-hart.com/resources/2003/unintended.htm. Accessed 24 Nov 2014.
58. Keefer, P., Loayza, N. V., & Soares, R. R. (2011). The development impact of the war on drugs. World Bank/World Bank/Pontificia Universidade Católica do Rio de Janeiro.
59. London Fire Brigade. Great Fire full story. http://www.london-fire.gov.uk/GreatFireFullStory.asp. Accessed 24 Nov 2014.
60. Merton, R. K. (1936). The unanticipated consequences of purposive social action. *American Sociological Review, 1*(6).
61. Roach, J. (2004). *Has mysterious killer of India's vultures been found? National Geographic News*. Washington: National Geographic Society.
62. Schofield, J. (2011). *London, after the Great Fire*. BBC., http://www.bbc.co.uk/history/british/civil_war_revolution/after_fire_01.shtml. Accessed 24 Nov 2014.
63. Smith, P. (2011). *Australia: derailment danger*. London: Financial Times. http://www.ft.com/intl/cms/s/2/f112a3ae-ce35-11e0-99ec-00144feabdc0.html#axzz3Ar0YFO6W. Accessed 24 Nov 2014.
64. Sterman, J. D. (2006). Learning from evidence in a complex world. *American Journal of Public Health, 96*(3), 505–514.

Chapter 10

65. Butler, T., Cane toads increasingly a problem in Australia. http://news.mongabay.com/2005/04/cane-toads-increasingly-a-problem-in-australia/. Accessed 9 May 2016.
66. Capra, F. (1997). *The web of life: a new scientific understanding of living systems*. New York: Anchor Books.
67. Peterson, S. (2010). *Systems thinking for anyone. Tracing connections*. Lebanon: isee systems.
68. Richmond, B. (2010). *The thinking in systems thinking. Tracing connections*. Lebanon: isee systems.
69. Sterman, J. D. (2006). Learning from Evidence in a Complex World, *American Journal of Public Health*.
70. Wikipedia, Deepwater Horizon oil spill. http://en.wikipedia.org/wiki/Deepwater_Horizon_oil_spill. Accessed 24 Nov 2014.

Epilogue

71. Breazeal, C. 5 ways robots will change your life in the future. http://computemidwest.com/news/5-five-ways-robots-will-become-personal-in-the-future/. Accessed 11 May 2016.
72. Creative Learning Exchange. http://www.clexchange.org/. Accessed 11 May 2016.
73. deCharms, R. C., Applications of real-time fMRI. http://www.wpi.edu/Images/CMS/HUA-CIMA/2008_deCharms.pdf. Accessed 11 May 2016.
74. Donald-Stanford, B. Girls 'rewire' brain to beat depression. http://www.futurity.org/girls-rewire-brain-to-beat-depression/. Accessed 11 May 2016.
75. Driverless car market watches: top misconceptions of autonomous cars and self-driving vehicles (2015). http://www.driverless-future.com/?page_id=774. Accessed 11 May 2016.
76. Gordon, T., & Mathias, L. (2014). *When will cars drive themselves?* Ingenia Online, http://www.ingenia.org.uk/Ingenia/Articles/927. Accessed 11 May 2016.
77. Gray, A. (2016). *The 10 skills you need to thrive in the Fourth Industrial Revolution*. Davos: World Economic Forum. http://www.weforum.org/agenda/2016/01/the-10-skills-you-need-to-thrive-in-the-fourth-industrial-revolution. Accessed 11 May 2016.
78. McLean, K. (2012). The healing art of meditation. Yale Scientific.
79. Mitchell, D., MIT Technology Review (2015) Household robots are here, but where are they going? http://www.technologyreview.com/news/537701/household-robots-are-here-but-where-are-they-going/. Accessed 11 May 2016.
80. Schwab, K. (2016). *The Fourth Industrial Revolution: what it means, how to respond World Economic Forum*. Davos, http://www.weforum.org/agenda/2016/01/the-fourth-industrial-revolution-what-it-means-and-how-to-respond. Accessed 11 May 2016.
81. Siemens website. Internet of things. http://www.siemens.com/innovation/apps/pof_microsite/_pof-fall-2012/_html_en/m2m.html. Accessed 24 Nov 2014.
82. UBS white paper (2016) On the Fourth Industrial Revolution; Davos 2016. https://www.ubs.com/global/en/about_ubs/follow_ubs/highlights/davos-2016.html. Accessed 11 May 2016.

Appendix I

83. Albertos, P., & Mareels, I. (2010). *Feedback and control for everyone*. Berlin: Springer.
84. Liptak, B. G. (2005). *Editor process control and optimization* (4th ed.). Boca Raton: Taylor & Francis.

Appendix II

85. Kim, D. H. (1992). *Toolbox: Guidelines for drawing causal loop diagrams. The systems thinker (Vol. 3, No. 1, pp. 5–6).* Westford: Pegasus Communications.

Appendix III

86. Bellinger, G. Archetypes: interaction structures of the universe. http://www.systems-thinking.org/arch/arch.htm. Accessed 24 Nov 2014.
87. Meadows, D. H. (2008). *Thinking in systems.* White River Junction: Chelsea Green Publishing.
88. Senge, P. (2006). *The fifth discipline.* New York: Currency Doubleday.
89. Wikipedia. System archetype. http://en.wikipedia.org/wiki/System_archetype. Accessed 24 Nov 2014.

Appendix VI

90. NetLogo. (2014). NetLogo user manual. http://ccl.northwestern.edu/netlogo/docs/. Accessed 24 Nov 2014.

Appendix VII

91. Schelling, T. (1978). *Micro-motives and macro-behavior.* New York: Norton.

Appendix VIII

92. IEC 61131–3. (2013). Programming languages standard for programmable controllers, International Electrotechnical Commission, Geneva, Switzerland.
93. IEC 61512–2, & ISA-88.00.02. (2001). Batch control standard part 2: data structures and guidelines for languages. International Electrotechnical Commission, Geneva, Switzerland and The International Society of Automation, Research Triangle Park, NC.

Index

A. Ghosh, *Dynamic Systems for Everyone*,
DOI 10.1007/978-3-319-43943-3